Praise

'Straightforward and practical financial information is essential for trades and construction businesses to thrive. *Know Your Numbers* is equipped with clear strategies that cut through financial complexity, making this an easy and relatable read. As the saying goes, "If you don't know your numbers, you don't know your business." This book provides refreshing insights that once applied will assist you in generating consistent profits, which ultimately create more financial freedom.'

— Jason Loft, Managing Director,
PROTRADE United

'This book is a must-read for ambitious business owners. Implementing the CLEAR method with our VCFO has seen our business grow five-fold and saved thousands of dollars in tax.'

— Tom Allen, Managing Director,
TGR Transport

'*Know Your Numbers* is a unique resource for the small to medium business owner. It will be confronting and concerning for some people. As business owners read the book, they'll be able to relate to the stories and examples and realise that their business is experiencing the same financial pitfalls and traps. Importantly, however, the CLEAR method is a simply explained process for overcoming these challenges and, with discipline, the business owner is given a model to follow that will guide their way towards success.'

— Michael Griffiths, Organisational and
  Leadership Development Consultant,
  Cornerstone Integral Solutions

'The need for this book is immense. *Know Your Numbers* is a gift to the reader. This is a must-read for all trades business owners who are looking to understand their numbers and succeed.'

— Kath Essing, published author, management
  consultant and leadership coach, Bespeak
  Consultancy

'Before reading this book, my team and I were stressed. We have implemented the CLEAR method, and now we know our numbers and are confident about our business finances and the direction we are heading.'

— Jim Davies, Managing Director,
  J & J Haulage

'The CLEAR method outlined in *Know Your Numbers* is exactly that. Its purpose is to help business owners take all the worry and uncertainty out of the important and time-consuming accounting function of their company. This lets them focus on their actual business activity, and ultimately allows them to live a good life as they work to build value in their business and see it grow through prudent financial practices and decisions.'

> — Peter Georgiadis, former Australian CEO of a leading global stainless steel manufacturing company

'The results we have achieved have been astronomical. The changes we have made to the business have been huge, and it makes me feel fantastic. *Know Your Numbers* is a must-read for all trades business owners.'

> — Rob Rossiello, Managing Director, Colourtech Coatings Pty Ltd

KNOW YOUR NUMBERS

The no-nonsense finance guide
to building a profitable and
enjoyable trades business

Lyndon Russell FIPA, FFA

R^ethink

To my family, for your unending support

Contents

Introduction

Welcome to a transformative financial journey for your trades business, a clear path designed to achieve 25/10 – elevating your earnings to $2,500 per week and securing a steady 10%-plus operating profit. This book isn't just a collection of ideas, it's a practical toolkit for turning ambition into attainable goals.

The statistics are sobering: 43% of small businesses fail to make a profit and 75% of small business owners take home less than the average wage.[1] The strategies presented in this book are designed to place you among those that thrive.

Financial missteps are often behind small business downfall. A National Australia Bank survey revealed a startling disconnect: only 19% of small business owners feel proficient in financial management, despite 58% recognising its critical role in success.[2] This gap is especially noticeable in trades businesses, as trade skill often doesn't equate to financial know-how.

Drawing from over two decades of experience and success stories like that of Geoff and Lyn, the dynamic duo behind a successful steel powder-coating factory who quadrupled profit in three months, this book is an actionable roadmap to direct you to financial visibility and monetary success. The businesses we admire – those that thrive without heavy marketing, consistently delight customers and can operate independently of their owners – share a common foundation: strong financial management. This not only ensures profitability, but also affords bosses the opportunity of life balance.

Guess what? Achieving success is not that hard.

This book caters to all trades businesses, from those starting out to seasoned veterans eyeing retirement. It covers your situation in two parts. In Part One, you'll discover the five biggest financial mistakes made by trades businesses and learn how to rectify them. You'll revisit the reasons you started in business and its significance in your world and delve into the importance of profit as a barometer of business health.

Part Two introduces the CLEAR method, a structured five-phase approach to achieve financial visibility, enjoyment and reward. In implementing this, you will travel along the CLEAR pathway:

- Clean-up
- Learn
- Enjoy
- Ambition
- Reward

Through this model, detailed case studies and expert advice, this book will guide you towards not just more substantial earnings, but a rewarding entrepreneurial journey. If you're aspiring to transform your business from constant struggle to a thriving profitable enterprise, this is your essential guide.

If you are ready to take your business to the next level, climb on board!

PART ONE
WHY YOU NEED THIS BOOK

ONE

Top Five Financial Mistakes Sabotaging Trades Businesses

H ave you ever wondered why so many trades businesses are flat-out, delivering a top-notch service but struggling financially? Over half of small business owners, defined as those employing fewer than twenty people,[3] wrestle with their own bookkeeping, despite 81% admitting a lack of financial acumen.[4] This is underscored by another sobering statistic: in 2016, 50% of small business owners in Australia registered a net income of less than $25,000 – less than the minimum wage.[5]

As an experienced accountant and advisor to countless businesses over two decades, I've pinpointed five major financial pitfalls that ensnare a struggling business:

1. DIY bookkeeping – doing your own books

2. Budget blues – a lack of planning

3. Breakeven blindspot – not knowing the point at which income covers expenses

4. Report neglect – staying uninformed

5. Tax planning failure – paying more tax than necessary

Not addressing these pitfalls places your business at risk of being among the 82% of failures caused by poor financial management.[6] Fortunately, pitfalls can be overcome. By the end of this chapter, you'll understand how businesses like Geoff and Lyn's turned their financial shortcomings into stepping stones for success.

Mistake #1: DIY bookkeeping

There are many reasons why business owners handle their own bookkeeping. Maybe at the start, it kept costs low and was fine when the workload was light – say at fifteen transactions per week. However, when the business expanded and things like extra staff, new loans, leases and additional expenses kicked in, the complexity demanded higher skill levels.

Imagine you are the proprietor of a growing landscaping business. You've scaled up, adding two new team members and investing $90,000 in a tipper truck, financed over four years. This expansion increases total monthly outgoings by $14,000 ($11,500 per month for wages, plus $2,500 for operational cost and loan interest on your truck).

That's not all. With the hiring of the two staff members, you need to account for oncosts such as superannuation, work cover, Christmas bonus, annual leave, sick leave, long-service leave, as well as training and management costs.

If this sounds familiar to you, I ask: 'How's your DIY bookkeeping coping with all these extras?'

As your business evolution continues, your bookkeeping needs to keep pace. Persisting with a DIY system means failing to address rising financial requirements. You must adapt to the business's shifting needs and ensure its bookkeeping proficiency matches the increasing complexity.

In short, bookkeeping is the lifeblood in understanding business finances. With good quality bookkeeping comes knowledge and clarity of the important numbers which enable sound financial decision making. Only then comes sustainable business growth.

The common problems encountered with wonky DIY bookkeeping are:

- **Software skills deficit.** Being proficient across bookkeeping software necessitates professional training. Lacking this expertise leads to critical errors or omissions, which reduce financial visibility.

- **Inadequate bookkeeping experience.** As the volume and complexity of transactions grow, those without adequate experience falter. Possessing professional bookkeeping skills is a must for navigating business expansion.

- **Convoluted chart of accounts.** A lengthy chart of accounts creates confusion. For example, a client had thirty-two different accounts for vehicle expenses, with each van having separate accounts for fuel, registration, insurance, repairs, tolls, etc. This made the profit-and-loss report overwhelmingly detailed.

- **Risk of incorrect goods and service tax (GST) categorisation.** Misunderstanding GST categories will lead inexperienced DIY bookkeepers into a maze of confusion. Selecting wrong GST codes in bookkeeping software not only results in flawed business activity statements, but also heightens the likelihood of compliance errors and dreaded GST tax audits.

- **Misclassifying owner payments.** Business owners extract money from their business, but mislabelling withdrawals as 'wages' or 'expenses' skews the profit-and-loss report. It's critical to identify owner payments correctly into either wages, loans or drawings.

- **Unreconciled bank and credit-card statements.** Contrary to software companies' claims of easy one-click reconciliation, the reality is far more challenging. True, most transactions are easy to reconcile, but others require skill and expertise. Neglecting to resolve trickier transactions will spawn inaccuracy in your financial data and reports.

- **Inadequate receipt and record-keeping.** If mistakes are made, you'll need to check and correct the original tax invoice. Without proper record-keeping, this becomes impossible.

- **Infrequent review of financial reports.** Poor knowledge of the software, an unworkable chart of accounts, unreconciled transactions and uncertain GST records all lead to a lack of trust in your financial reports and create significant doubt in your decision making.

Mistake #2: Budget blues

A 2022 Clutch survey revealed only 50% of businesses prepared a formal budget in 2020.[7] It showed that the smallest businesses, those with ten or fewer employees, were the worst offenders, with only 19% preparing one. By contrast, 100% of larger businesses with 250+ employees had a formalised budget.

These findings highlight the self-inflicted financial handicap that small business owners place themselves in by failing to prepare a formal budget. Without a budget, small businesses risk overspending and underinvesting, which ultimately undermines their financial viability.

Businesses that have a budget are better equipped to manage their resources effectively, which helps prevent overspending and underinvestment. As a result, they are well placed to keep finances optimised for performance – the launch pad for growth.

Formal budgets are essential, so why don't small businesses budget? Good question. Here are some of the excuses I've heard over the years:

- 'I want to focus on running my small business and doing what I love. I don't want to stuff around with business budgets. I don't understand them.'

- 'I just look at the balance in my bank account, and it seems to be growing every month, so why do I need a budget?'

- 'I'm too busy with the day-to-day operations of my business to worry about creating a budget.'

- 'I tried to make a budget once, but it was too hard.'

When financial reality comes home to roost and the excuses don't stand up, these same people say:

- 'I have to pay *what* in taxes?'

- 'I don't have the money right now.'

- 'I'm making money, but it's not showing up in my bank account. I made a profit, so where is it?'

- 'When I first started my business, I had money, but as the business has grown, expenses have spiralled out of control.'

- 'Money's coming in, but it seems to be going out the door faster.'

- 'I don't know how to put the matter right.'

Why should small businesses budget?

According to the University of Technology Sydney, 32% of small business collapses are due to poor planning and financial management.[8] The statistics are clear: brushing budgeting aside is a direct path to failure.

Imagine driving to an unfamiliar destination without a Global Positioning System (GPS) or map. When you reach the intersection, you won't know whether to turn left, right or go straight ahead, because you don't know which way you are heading. Just as you wouldn't hit the road without direction, you shouldn't operate your business without a budget.

Businesses that fail to budget show all the common symptoms of financial stress, such as cash-flow crises, unfeasible margins, reckless credit use and lack of a financial plan. Preparing the formal budget is a road-map to steer a business towards viability and away from disaster. In other words, it gives the direction to reach realistic goals.

CASE STUDY: Geoff and Lyn's first budget

Meet Geoff and Lyn, the dynamic duo behind a successful steel powder-coating factory. Leading a dedicated team of eight at their expansive facility, they transform raw metal in their enormous oven, creating high-quality corrosion-resistant coloured-steel products.

Geoff, a hands-on man, initially believed that sheer hard work was the key to making money, but an analysis of the business's financial history told a different story. The profit in the previous year was $128,000 before he and Lyn paid themselves a wage. After taking into account a wage of $130,000 for Geoff and $65,000 for Lyn, the business was operating at a loss of $67,000. Financial disaster was looming.

In consultation with Geoff and Lyn, I prepared a budget which factored in sales revenue predictions, profit margins, staff costs, rent and overheads in aspiring to a 10% operating profit. The preparation and analysis of their first budget revealed gross profit had to increase by 8% to reach the mark. The budget showed the way.

Through monthly meetings, I provided the actual results relative to budget, which allowed Geoff and

Lyn to make necessary adjustments to hit target. The result? The first year saw their business rake in a solid $163,000 profit, after paying a commercial wage to Geoff and Lyn. For the first time in twenty years, they made real money.

Flush with cash, Geoff could take a detached view of his business and the numbers. When I asked him how he felt about his achievement, he said, 'Take it as gospel, it's empowering to go from believin' you're doin' it right, to knowin' you are.'

Here lie the perils of operating without a budget:

- **Risk of underpaying yourself.** With no solid budget, you might be steering a busy business while grappling with inadequate take-home pay (Geoff and Lyn's initial predicament).

- **Crippling financial chaos.** Budget-less businesses walk a tightrope. They risk not fulfilling monetary commitments to staff, suppliers, lenders and the tax office. This approach can trigger a financial freefall which leads to disaster.

- **Crisis of financial direction.** No budget equates to navigating without a compass. If you lack financial controls provided by a budget, overspending usually follows. Decision making becomes guesswork and financial stability balances on a knife edge.

> **PRO TIP**
>
> Professional bookkeeping streamlines budget preparation. A convoluted chart of accounts with unreconciled transactions and incorrect allocations makes budget preparation an uphill battle.

Mistake #3: Breakeven blindspot

Knowing your breakeven point is key to operating a successful business. This is the point at which total revenue equals total expenses. Without a clear understanding of this point, businesses struggle to determine the level of sales revenue and production capacity required to cover costs and start making a profit.

Furthermore, understanding the breakeven point offers businesses a valuable tool for assessing their performance. By comparing actual sales to the breakeven point, business owners can gain clarity and focus on their real-time profit or loss position. This information can then be used to make better decisions about pricing and production levels to achieve a profitable outcome.

Knowing and monitoring the breakeven point provides entrepreneurs with profitability intel. This information helps them to make decisions that will improve profitability and overall success.

In a nutshell, knowing your breakeven point eradicates uncertainty and simplifies business management. Better outcomes will be shown through:

- **Unleashed potential.** Knowing your breakeven point provides clarity and focus. You know the minimum revenue needed to meet all expenses, enabling you to set achievable targets and readily assess business profitability.

- **Enhanced accountability.** Tracking your breakeven point enables better financial control. Not reaching breakeven will prompt further investigation. If you check monthly, you can identify a sudden financial shift and act swiftly.

- **Empowered decision making.** Knowing your breakeven point elevates decision making, as you have a concrete benchmark against which to assess the impact of business decisions. For instance, the decision to expand your team increases the monthly breakeven point due to higher costs. You can immediately calculate how much extra revenue needs to be generated to avoid losses.

- **Mastered costs.** Identifying your breakeven point will bring on examination of costs. You will ask, 'What costs can I eliminate to lower my breakeven point?' This insight empowers you to manage business expenses better and optimise financial performance.

Mistake #4: Report neglect

Regular reviewing of financial reports is essential because it allows businesses to track financial performance, identify areas of improvement and make informed decisions. By reviewing financial reports, business owners can assess their cash flow, monitor expenses and evaluate revenue trends. This helps in identifying any potential issues early on and taking corrective actions.

Additionally, regular financial reviews enable businesses to assess their financial health, identify strengths and weaknesses, and make plans for future growth and sustainability. Overall, a regular review of financial reports provides valuable insights and helps in ensuring sound financial management for small businesses.

Suppose you are reviewing your financial reports and discover a $9,000 expense under 'printing and stationery', an expense category that typically costs around $200 per month. Upon enquiry, you find it's for a photocopy machine – an asset, not an expense. This oversight, left uncorrected, will distort your profit-and-loss statement, inflating expenses, diminishing profits and creating the illusion your business is underperforming.

Regular review of financial reports is non-negotiable. It's not only about catching errors, but also steering your business in the right direction.

Mistake #5: Tax planning failure

The fifth big mistake is not tax planning. Before your eyes glaze over, let me remind you that the late great Kerry Packer, one of Australia's richest men, was once grilled by the Senate Estimates committee about his taxes.[9] He told them, of course he was minimising his taxes, within the boundaries of the law, and if anyone in the country was not trying to do the same, they needed their heads read. He went on to add that the government was not spending its money that well, so he didn't want to donate any extra.

Tax planning is not a conversation everybody enjoys at a Sunday barbecue, but it's a huge potential windfall for small businesses. Take, for example, a business with an annual revenue of $2–3 million. Strategic tax planning might save $20,000 or more each year – a sizable sum that could fuel business growth. With that kind of money, you could revamp your website, upgrade plant and equipment, or expand your team. All worth doing.

It's not about tax return numbers – but rather, accumulating tax savings that over a decade could amount to hundreds of thousands. Big dollars.

CASE STUDY: Geoff and Lyn's tax strategy

Ready for a fun fact? Geoff and Lyn reduced their tax bill by $52,000 in a single year!

How did they do this? With a tax-effective investment in a second oven valued at $200,000.

What were the circumstances? Previously, small powder-coating jobs would take a back seat when the large ones were baking. This led to costly production delays. A smaller oven, they felt, was the solution, but the price was hefty. As they were true entrepreneurs, they took the plunge.

It paid off! The $200,000 cost was accounted for as a plant and equipment asset on the balance sheet, and under tax law, Geoff and Lyn utilised the Temporary Full Expensing provisions to instantly claim a full tax deduction for the asset purchase. This reduced the taxable profit by $200,000, thereby saving $52,000 in company tax. Not too shabby – the tax office funded 26% of the asset cost.

Wait, there's more! Not only were the tax savings significant, but the installation of a smaller oven increased production capacity and sales. Monthly sales, pre-second oven, were around $170,000, but after became $205,000 – a 20% revenue increase. That's the power of tax planning.

Summary

According to accountancy experts Nine Advisory, 60% of small business owners feel overwhelmed and lack sufficient knowledge about the accounting and finances of their business.[10] Yet so many have a DIY approach to bookkeeping. If that's you, I say, face this challenge, give it immediate attention. Get a book-keeping pro today.

Why a professional? Because bookkeeping is the lifeblood of your business finances. If you are serious about profits and knowing your numbers, you need a pro. They allow you to level up your business finances, making it easy to design a budget to deliver 25/10.

When you understand and adopt the budget, the financial fog begins to lift, allowing you to identify your monthly breakeven point accurately. Knowledge of the breakeven point means you can set revenue and production targets relative to the financial goals. This provides knowledge of the numbers you need to be shooting for.

With all the financial tools in place, your trust in your financial reports will rise. You will look forward to reviewing them monthly relative to budget and breakeven.

Having a keen awareness of financial performance will naturally lead to you considering taxes and the amount that needs to be paid. Accessing quality tax planning services can help you save your business money by reducing the tax burden. The money saved can then be reinvested into further business growth.

The ultimate goal of this journey is not just to equip you with financial knowledge – it's to arm you with acumen and pro-level insights to drive your business towards exceptional financial prosperity.

TWO

The Importance Of Your Business

In the busyness of life, business planning often takes a back seat. Although you started the business with a clear vision to secure a prosperous future, as time passed, amid the hard work and endless hours, that vision seemed to fade. Your efforts sometimes feel like they yield little reward, and the struggle becomes all too real.

Your business is important. Why? Because it's the only ticket to high earnings and a dream lifestyle. There is boundless potential locked up in your business. You just have to use the right key to unlock that potential for higher-quality leisure time and work-life balance.

Your success positively impacts everyone connected to you. It's time to rekindle that initial spark and align your business strategies with your personal goals. Remember, there's a community of stakeholders cheering for you, supporting your aspirations.

Consider the people whose futures hinge on your business's success:

- **Your family.** If you have a family, your children rely on you and your partner for financial stability. The relentless flow of mortgage payments and bills is your responsibility. Beyond financial support, your family looks up to you as the rock, their steadfast support. As a parent and role model, you have to be a beacon of strength. You're balancing business commitments with family time. A demanding gig.

- **Customers.** Your customers count on you. They trust in your ability to deliver on schedule, meet every deadline and exceed all expectations with a top-notch service.

- **Team members.** Your team depends on you, not just for salaries and benefits, but for a secure and safe work environment. They look to you for growth opportunities, skills development and a supportive work culture. When you look after them, you are positively impacting their families and wider community.

- **Subcontractors and suppliers.** To subcontractors and suppliers, you are a big cog in their world. Their livelihood hinges on you paying bills on time, adhering to schedules, maintaining clear communication and ensuring your part of the project sails smoothly. They're not just associates; they're partners in your journey to success.

- **Your community.** Your business drives part of the local community. Every employee you hire, supplier you work with and customer you serve boosts the region's economy. Beyond this, you might sponsor the local football team or rejuvenate the community centre. Your actions contribute significantly to the community's wellbeing. Your business is more than a local entity; it's an essential part in building a stronger, more vibrant Australia.

- **You.** Last but not least, you depend on your success. Your emotional and financial wellbeing are tied to your ability to balance all dependencies while operating a successful business. One thing for sure, an unsuccessful business makes you unhappy. You cannot achieve success if you are overwhelmed by stress, fatigue and financial insecurity.

Projects in trade and construction are worth big bucks – tens of thousands, hundreds of thousands or even millions. Yet in 2022, the Association of

Professional Builders stated that around 50% of construction companies in Australia were experiencing negative equity and around 25–30% were unable to pay their bills on time.[11] This highlights a hidden risk: projects may fail because of the weak financial health of construction companies.

Furthermore, the collapse of the Melbourne-based construction giant Porter Davis highlights the severe repercussions. In 2023, the *Sydney Morning Herald* reported that this collapse resulted in the halting of construction on 1,700 homes.[12] Thousands of customers were left with either an unfinished house or an unrecoverable deposit. Heartbreaking for everyone affected.

These statistics and news stories of collapsing construction companies show that the stakes are high when customers work with unprofitable, financially struggling businesses. Therefore, it's vital your business be profitable so it can deliver the products and services your customers pay for.

You're obligated to yourself and your stakeholders to make your business profitable, sustainable and resilient. Your business's financial success must be your top priority.

Let's refocus and navigate this path together, turning your business into the powerhouse of your dreams. We'll now look at how to do this.

Your goals

Let's go back to the beginning. What were your reasons for starting your business? Remembering these motives will provide clarity.

Some common motivations for starting a business include:

- Increasing income
- Pursuing a passion
- Escaping the corporate rat race
- Discovering your purpose
- Being self-employed
- Having control over your schedule
- Learning new skills
- Having comparative freedom
- Increasing job satisfaction
- Building a team
- Creating a brand
- Giving back to the community
- Embracing the challenge

I've spoken with hundreds of business owners in trades and construction over the last twenty years, and if you're anything like them, this might be your story. You are a skilled tradesperson but have higher ambition.

Let me introduce you to Mick.

CASE STUDY: Mick's first turning point

Mick, a forty-five-year-old carpenter, now operates his own building and construction company specialising in custom homes. His business utilises Buildxact software and employs twelve full-time staff at his factory, located in an industrial estate in a large town in regional Victoria. Mick is married and has two teenage daughters attending private school. He has a mortgage on a comfortable home, much of which he built himself.

At school, Mick wasn't academic, but loved working with his hands and being outdoors. His calling was carpentry, and after a local apprenticeship, he seized an opportunity to join a volume builder as leading hand.

Mick was good at what he did and committed to working hard and producing quality work. Things were great – he married his partner and was promoted in his job. By age thirty-three, he and his wife had two daughters and he was a project manager.

Although successful, he questioned himself. What did he want in his career? He wanted to stretch his skill level. What should he do about it? This took some thought.

This thinking led him to re-evaluate his personal goals, his first turning point. With a pen and paper in hand, he wrote down the three goals he wanted to achieve by running his own business:

- **Financial security.** Mick wished to elevate his family's financial position. This would mean a better lifestyle for his wife Beccy and improved education opportunities for his kids. He understood the necessity for a solid business plan.
- **Control.** Mick wanted to control his future – to be his own boss and work for himself. He loved the idea of being master of his own destiny.
- **Job satisfaction.** Mick wanted to produce high-quality custom homes, which in turn would deliver higher value to those whose homes he was building.

When you're starting a business, your motives are crystal clear. However, as time progresses, the complexity of operations inevitably grows. The customer base expands, the network of subcontractors and suppliers broadens, the team enlarges, and you're navigating through a maze of equipment loans, taxes and compliance obligations. As business evolves, demands intensify.

In response, you turn to software solutions to manage this escalation, suddenly finding yourself juggling timesheets, scheduling, material procurement and invoicing. Amid this whirlwind of activities, the reasons why you originally started your business blur. You're in a relentless race to stay ahead. Each day becomes a balancing act; fall and you fail.

Most owners of trades businesses started their careers as apprentices before learning their skill as tradespeople. The step from there to business operator is big. Each one had to make a leap of faith from good tradesperson to even better business operator, and these skills are not taught at trades school.

Cash-flow management is one of the skills not taught. Therefore, all you see is money coming in and going out – the revolving door. Management accounting is not taught either. All you see is the profit-and-loss report, but how sure are you of the profitability?

This is no one's fault, it is circumstance.

To change your circumstance, look outside and engage specialists with skills and abilities to help in the areas where you fall short. When you inject specialist skills into your business, mastery of business finances is in sight.

Why don't owners of trades businesses get the help they need to meet their goals? There are three main reasons:

- Ignorance
- Denial
- Plans and goals have not been identified

Some clients come to my accountancy business, Next Level, without having goals in place. From the hundreds of conversations I've had with these individuals seeking help, I've found the reason for this is often that they hold self-limiting beliefs.

Let's look at what I call the belief set. What are beliefs? There are two types: limiting and expanding. We'll look at both.

Limiting beliefs

A limiting belief is a self-image and view of the world at large which may hold people back from taking opportunities, reaching their potential or achieving their goals. I reckon eight out of ten business owners facing financial struggle hold either limiting beliefs or financial ignorance.

Here are the top six limiting beliefs I've heard from trades business owners over the years:

- Raising prices is impossible; my customers will never accept it.

- I must do this high-value job myself. What if my team makes a mistake?

- I'll cut costs by doing the work myself.

- Working harder guarantees success.

- I need to find an easier way to make money – I'll look out for an exciting idea for a new product or service.

- As long as there's work at the door, cash flows in.

Limiting beliefs are characteristic of an owner who is working in the business rather than on it. Limiting beliefs can trap business owners in a cycle of struggle for years, even decades. If this sound like you, change it. Put yourself first and reset your thinking towards expanding beliefs.

PRO TIP

Why does airline safety advice recommend securing your oxygen mask before assisting others? By putting on your own mask first, you ensure your safety, and are then able to help others without putting yourself at risk. The same goes for the financial health of your business. Fix that first, and you are then able to help others (key stakeholders) without putting your business finances at risk.

Expanding beliefs

An expanding belief is a positive mindset that encourages personal growth, development and openness to new opportunities. Unlike limiting beliefs, which can hold you back, expanding beliefs empower you to reach your full potential and achieve goals.

Define your goals, ink them down, make them real. Commit to your plan, ditch limiting beliefs and embrace new horizons. Don't survive, thrive.

With this mindset, you will see things differently. You'll think and say things like:

- Raising prices is possible; my customers will see value.

- My team can handle this high-value job; mistakes, if any, are opportunities for learning.

- The business will make more money if I delegate.

- To encourage success, I need broader opinions and the right advice to find the way.

Moving from a limiting to an expanding belief set involves a change in mindset. Once you have the right mindset, the sky's the limit. Part Two of this book will explain the CLEAR method, a proven five-step process designed to elevate your thinking and propel you towards high levels of success.

A GIFT OF $50,000

What would you do if you were given $50,000 to invest in your business? How would you spend it?

This is a great hypothetical question. It eliminates the concern of using your own money, encouraging expansive thinking about business investments without any limitations. This opens up the possibility of pursuing new options that have the potential for high returns. It allows you to think outside the box and consider additional ideas that could lead to significant growth and success.

Goal reset

Nothing succeeds like success – the secret of success is well explained in this one little phrase. It means that if you are successful in one thing, then you are more likely to be successful in more things in the future. Apply this to your own business, be successful.

Why do people with goals succeed? Because they know where they are going.

Setting goals in business is a mindset for success. It provides a clear roadmap that everyone can follow and work towards. Goals enable businesses to measure progress against a target and identify areas for improvement, helping create a high-performance culture.

In his bestselling book *The E-Myth Revisited*,[13] Michael Gerber talks about the difference between successful people and everyone else. He says successful people actively create their lives rather than passively waiting for the next thing to happen. The message is clear: create a goal for your business. Enjoyment will come in the pursuit of working towards and achieving it.

Let's do a goal reset exercise. What were the personal goals you had upon starting your business? Were they any or all of these?

- To pick up the kids from school daily.
- To take my family on a nice holiday every year.

- To have more time with family, friends and loved ones.

- To become financially independent.

Re-evaluate whether you are on track towards achieving your goals or drifting away from them. If you've strayed, recalibrate your focus and make those goals a priority once again. It's never too late to refocus and reignite that initial spark that drove you to start your business.

The best advice I have for you is to think big. Resetting your goals, particularly big, audacious ones, can help you break free from the rut you may find yourself in.

CASE STUDY: Mick's second turning point

Mick's journey is an inspiring tale of grit and determination. He started his business with a dream and a passion. Over a decade, he brought in impressive projects, expanded his operations and even owned a factory. Yet, as his business flourished, Mick found himself trapped in a paradox of success.

He was working tirelessly, clocking in sixty-hour weeks, but his income remained stagnant at $80,000. His reliable staff, ironically, were earning more while working fewer hours. His wife couldn't help but question the imbalance.

Mick had achieved his goal of delivering a quality product, but at a cost. His job satisfaction was dwindling and financial security was a fading dream.

The wake-up call came when his younger daughter started private high school, doubling the school fees. Managing with one child at this school was a struggle, but two was a breaking point. Mick knew he had to change course.

Recognising his limitations, Mick sought business financial advice. He was a master at building houses, but managing finances was a different ball game. He had an accountant for basic services, but he needed more. He required strategic guidance at the chief finance officer (CFO) level to navigate a path towards his financial objectives.

Seeking help marked the second turning point in Mick's career. That's when he spoke to the team at Next Level. We sat down with Mick and discussed how we'd work together to achieve 25/10.

Mick smiled and said, 'Beccy will be so happy, this is exactly what we want.'

What 25/10 meant for Mick and Beccy was:

- More take-home pay
- Ability to pay the private school fees for two kids
- Financial security

To achieve this, we took these steps:

- **Set a new income goal**. Mick began earning $130,000, adding $50,000 to his previous income. This allowed him to pay his daughters' private school fees upfront.

- **Improved business profitability**. We created a budget that accounted for Mick's salary and other costs, incorporating a 10% operating profit goal.

- **Refined pricing strategy**. Mick hadn't raised his prices in four years and they were below market standards. We highlighted the areas where he could adjust his pricing.

By implementing these strategies, Mick had the game plan to prioritise his earnings and achieve a 10% operating profit in his business – 25/10.

Common fears

Despite having a plan, many trades and construction business owners still find it difficult to overcome their limiting beliefs. Here are two widespread fears these owners have.

My customers won't accept a price increase

When business owners face self-doubt about repricing their services, they tend to ask themselves:

- What if customers don't accept it?

- How many clients would I lose as a result?

- Would I lose all my clients?

- Would I need to lay off staff?

- Would I need to sell some of the equipment in the business?

- Would I need to downsize?

Why do these doubts creep in? Answer – limiting beliefs.

> **PRO TIP**
>
> When quoting a job, don't rely on what your mates with similar businesses charge. Who says their pricing is correct? Determine your own prices.

If the goal is to have more free time, your business needs to become more profitable to afford the investments you'll require in staff and systems. If the goal is to increase your take-home pay, the business needs to become more profitable to afford to pay you what you deserve. The goals you've set drive your motivation.

A price hike is a rebalance to accurately reflect value provided. Value refers to what the customer receives, while price refers to what they pay. Price and value must align equally. The equation becomes misaligned when prices are not reviewed regularly.

> **PRO TIP: Value over price**
>
> Putting your prices up doesn't mean you're ripping off customers. If you charge a fair amount for the value you provide, nobody is getting ripped off. Ultimately, the customer is happy to pay the price for the value they receive.

When it comes to resetting prices, business owners fear they may lose 50% of their customers. However, in my experience, they don't lose more than 5%, and those are often a pain anyway! They're the customers who are so price-focused that every time you send them a bill, they pay late or dispute it, tying you up in justifying things. These are not the kind of customers who see the value in what you have created for them.

I've helped hundreds of business owners to set budgets and create goals. When the plan says a modest price rebalance is required, much of the fear associated with taking this action dissipates. When the price reset has been completed, it's often a breakthrough moment in the journey towards achieving your dreams.

> **PRO TIP: Attract the right customers**
>
> Low rates attract customers who are cost conscious, not value conscious. When you run a niche business, it's important to attract customers who value your services.

My staff can't handle the tricky jobs

I've sometimes arrived at a client's business to find them out on the factory floor, goggles and gloves on, working on a machine. When I ask, 'Why are you working on the tools today?' the answer is often, 'Oh, because it's a tricky bit of work and the team can't do it.'

That's going to be true in many different areas of your business. Chances are, you started the business because you excel at what you do. However, you need to build your confidence in getting your staff to a point where they can do the job. Maybe they won't be 100% as good as you, but maybe they'll make it to 70%, and next month they'll reach 72%, and the next month 74%.

You've got to let go of the fear of imperfection. This is the only way you can develop your team, otherwise you will become a bottleneck in your business. If you're the only one who can do everything, guess what? You're going to end up doing everything. You've got to be comfortable with your staff doing the tasks to a 70% standard, and then work towards getting them to 80%.

An example is the iPhone – Apple didn't wait until it had nailed the technology to the standard we have now. No, it released the iPhone back in 2007 with 2007 technology, and then set about improving it every year. If you are always waiting for perfection, you'll never do anything. It's better to start something, and then improve on it.

Good customers mean good business

What makes one customer better than another? Answering this question can help you focus on the customers you want to attract and make it easier to let others go when they are not worth your while.

Perhaps you like to work with clients who:

- Focus more on the value you provide rather than the price you charge

- Value everything you do for them

- Don't complain about fees and charges

- Pay their invoices on time

- Understand and accept contract variations and the associated fees

- Have everything ready so that your business can get on with the job

- Communicate well

It's a pleasure doing business with customers like these, and guess what? The best customers want to work with the best businesses. Businesses that:

- Have clear and transparent pricing systems and methods

- Are highly profitable

- Can clearly articulate the value they provide

- Are well organised and schedule all stages of the project efficiently

- Ensure trade service and materials turn up on time

- Deliver the job on time

- Deliver the job on budget

- Complete the work with minimal defects

- Complete the work to a high standard of quality

- Provide great customer service

- Communicate well

If you consistently demonstrate the high value you provide through the work you do, the interactions you have with your customers, and the reliability, competence and skills of your team, you will automatically attract the customers who actively want to work with you. These customers won't argue and moan about the price, because they will appreciate and have confidence in the high standards they will receive from you. Then you will wonder why you ever worried about losing those pesky price-conscious customers.

Summary

The first step in untangling your financial woes is to go back to your original goals. When you started your business, you had a clear vision, but somewhere along the way, it likely became blurry. As the business got busier, you had to work harder to keep up.

Before you know it, you've forgotten all about your goals. If this sounds like you, a goal reset is required.

Sometimes your goals will stay the same as when you started. At other times, when resetting goals, you'll find that you have new aspirations for your business. For example, remember Geoff and Lyn? They had been working hard in their steel powder-coating business for twenty years and were in their late-fifties. When they came to Next Level and we did a goal reset, we worked out that one of their new goals was to have an exit plan that would allow them to retire in five years' time.

Take this opportunity to write down the top three goals you have for your business. Here's another question – how much do you want to earn? What is a commercial wage for your role in your business?

Think big and have fun answering these questions. However, if you aren't able to answer them, don't despair. One of the first steps in the CLEAR method, described in Part Two, is to lay down a framework to clarify your vision when everything's murky.

It can sound a little theoretical when someone says, 'You've got to increase your prices' or 'You have to train your staff.' It's all just an idea, but when you use the Next Level CLEAR method, these ideas go from undefined to blueprint.

At this point, my clients often say, 'You know what, I can see what you're saying here and it's beyond doubt. It's all laid out in the figures and I am going to take the next step because it's in black and white in front of me.' This can be a breakthrough moment for them – and one of my favourite parts of my job.

THREE

Profit Is Not
A Four-Letter Word

P rofit is the reward that a business earns for providing value to its customers. While the customer does not intentionally award profit, they recognise value offered and pay what they consider a fair price.

Many trades and construction businesses begin with a fair balance between profit and value, but as knowledge grows and develops within the business, so do skills. The owner hires more staff, buys better equipment, establishes better processes and provides a higher-quality service. Raising prices is inevitable.

In this chapter, we'll explore the critical importance of setting the correct profit and pricing mechanisms in your business.

Vanity vs sanity

Have you heard the business saying, 'revenue for vanity, profit for sanity'? This means that while revenue may make a company look successful, it's the profit that really matters for its financial health.

CASE STUDY: A story about John and Sarah

Once upon a time, there were two entrepreneurs, John and Sarah, who started their own companies in the construction industry. John focused on generating high revenue and expanding his business rapidly. He invested heavily in marketing, hired a large team and pursued every opportunity for growth.

Sarah took a different approach. She prioritised profitability and carefully managed her expenses. She analysed her costs, optimised her operations and made strategic decisions to ensure a healthy profit margin.

A few years later, both John and Sarah faced a challenging economic downturn. Many of their competitors struggled to survive, but John's company was hit hard. Despite its impressive revenue, the business accumulated significant debt and had high fixed costs that couldn't be sustained during tough times.

In contrast, Sarah's company weathered the storm OK. Sarah minimised costs without sacrificing opportunity, which kept the company financially stable. Due to her focus on profitability, the business had built up cash reserves and had the flexibility to adapt.

This case study highlights the importance of prioritising profit over revenue. Remember, it's not just about how much money flows into your business, but how much you can keep as profit.

CASE STUDY: Peter and Brett's profit sanity approach

Peter and Brett, two long-time friends in their early forties, run a successful plumbing business that experienced significant revenue growth. Over three years, their turnover increased from $1 million to $3.5 million, which allowed them to expand their team to eight full-time employees and multiple subcontractors. Revenue was rolling through the front door.

However, with rapid growth, financial commitments increased, and Peter and Brett faced a challenge managing cash flow and profitability. When they received overdue payment reminders from the Australian Taxation Office and their biggest supplier in the same week, they realised the necessity for a rethink.

Seeking guidance, they contacted my team at Next Level. Immediately, we helped Peter and Brett establish a budget, develop a cash-flow forecast and implement tighter financial controls to align performance to the budget and cash-flow plan.

At our first virtual chief financial officer (VCFO) meeting, we talked figures: the ones they were aware of, the ones they weren't aware of and others of significant importance. Over time, as they worked closely with Next Level, Peter and Brett transitioned from feeling uncertain about the right questions to ask, to becoming so well-versed in their business's financial figures that asking became understanding. It was at that point they knew how to drive their business forward. That is, their knowledge allowed them to address business issues early.

Today, Peter and Brett are top performers in their field. Their focus on profit paid off with an optimised operating profit margin exceeding 10%. They now enjoy financial stability and predictable profitable results. Always looking ahead, they are excited about future projects and their continued success.

Types of profit

So that you can begin to understand your numbers, here's a refresher on key concepts. The definition of 'profit' is revenue minus expenses, but there are three types of profit.

Gross profit

This is the profit a company makes after deducting the cost of goods sold (COGS) from its total revenue. It represents the amount of money left over after accounting for the direct costs associated with producing or delivering a product or service. Gross profit indicates the profitability of a company's core operations before considering other expenses such as overhead costs and taxes.

$$\text{total revenue} - \text{COGS} = \text{gross profit}$$

Example:

$$\$1,000,000 \text{ (revenue)} - \$350,000 \text{ (COGS)} = \$650,000 \text{ (gross profit)}$$

Gross profit margin (GPM)

This represents the percentage of revenue that remains as gross profit after deducting the COGS. It is calculated by dividing the gross profit by the total revenue, and then multiplying that figure by 100. GPM indicates the profitability of core operations and is used to assess how efficiently a company generates profit from its direct production or delivery costs.

$$\text{(gross profit} \div \text{total revenue)} \times 100 = \text{GPM}$$

Example:

$$(\$650,000 \text{ (gross profit)} \div \$1,000,000 \text{ (revenue)}) \times 100 = 65\%$$

PRO TIP

GPM is crucial. It's your financial north star, your guiding light. It should be stable and growing. Businesses that offer more value than others tend to have higher GPMs. Focus on lifting yours.

Operating profit

This is generated from core operations after deducting expenses such as salaries, rent, utilities and other costs directly related to running the business. It measures a company's profitability before considering factors outside of its day-to-day operations, so excludes non-operating income or expenses like interest and taxes.

$$\text{total revenue} - \text{COGS} - \text{operating expenses} = \text{operating profit}$$

Example:

$1,000,000 (revenue) – $350,000 (COGS) – $500,000 (operating expenses) = $150,000 (operating profit)

Operating profit margin

This covers the percentage of revenue that remains as operating profit. It is calculated by dividing the operating profit by the total revenue, and then multiplying that figure by 100. Operating profit margin measures the profitability of core operations by indicating how efficiently the company generates profit from its day-to-day business activities.

$$(\text{operating profit} \div \text{total revenue}) \times 100 = \text{operating profit margin}$$

Example:

$150,000 (operating profit) ÷ $1,000,000 (total revenue)
× 100 = 15% (operating profit margin)

PRO TIP

When preparing a budget, prioritise the operating profit margin. If your business has an operating profit margin below 10%, it is essential to thoroughly investigate and identify the required measures to raise it to 10% or higher.

Net profit

Net profit is generated after deducting all expenses, including operating expenses, interest, taxes and other non-operating expenses. It is calculated by subtracting total expenses from total revenue. It reflects overall profitability and indicates how much money the company has left over after covering all obligations.

total revenue – COGS – operating expenses
– interest, tax, depreciation, amortisation = net profit

Example:

$1,000,000 (total revenue) – $350,000 (COGS)
– $500,000 (operating expenses) – $15,000 (interest)
– $10,000 (tax) – $65,000 (depreciation)
– $5,000 (amortisation) = $55,000 (net profit)

Net profit margin

This is a financial metric that represents the percentage of revenue remaining as net profit. It is calculated by dividing the net profit by the total revenue, and then multiplying that figure by 100. Net profit margin measures the profitability of a company's overall operations and indicates how efficiently it generates profit after covering all costs and obligations.

(net profit ÷ total revenue) × 100 = net profit margin

Example:

$55,000 (net profit) ÷ $1,000,000 (total revenue) × 100 = 5.5%

In summary, gross profit focuses on the direct costs associated with producing or delivering a product or service, operating profit includes operating expenses directly related to running the business, and net profit reflects the overall profit after deducting all expenses.

Other financial terms

While we're at it, here are the meanings for some other financial terms I'll be using:

- **Creditors** are individuals or entities to whom a company owes money or has a financial obligation. They include suppliers, lenders or any party that has provided goods or services to the company on credit. Creditors are typically listed as liabilities on a company's balance sheet and represent the amounts owed by the company to external parties.

- **Debtors** are individuals or entities who owe money or have a financial obligation to the company. They include customers, clients or any party that has received goods or services from the company on credit. Debtors are typically listed as assets on a company's balance sheet and represent the amounts owed to the company by external parties.

- **COGS** refers to direct costs associated with producing or delivering a product or service. It includes the costs of raw materials, of direct labour and directly involved in the production process. COGS is subtracted from the total revenue to calculate gross profit.

- **Revenue** refers to the total income generated by a company from normal business operations. It is the amount of money a company receives from selling products or services to customers. Revenue indicates the company's top-line performance and is typically reported on the income statement. It does not account for expenses or costs associated with running the business.

- **Discounts** refers to a reduction in the price of a product or service. A discount is often offered to customers as an incentive to encourage sales or as a promotional strategy. It can take various forms, such as a percentage off the original price, a fixed amount deduction, or a special offer or package deal. The purpose of offering a discount is to attract customers, increase sales volume and potentially gain a competitive advantage in the market.

- **Overheads** refers to ongoing expenses and costs necessary to operate a business, but they are not directly attributable to the production or delivery of goods or services. These expenses include rent or lease payments for office or production space, utilities, insurance, salaries of non-production staff, office supplies, marketing expenses and other general administrative costs. Overheads are essential for the day-to-day functioning of a business and are typically incurred regardless of the level of production or sales.

Profit is a five-letter word

Many trades businesses focus on delivering the highest level of workmanship, often at the expense of profit. It's almost as if profit is a four-letter word. However, profit is a five-letter word, and that word is great. Great, because profit is the primary function for the survival of your business.

A successful business operator recognises that profit fuels the business's development by providing the cash required for investment. It is important for owners of trade businesses to shift their mindset and acknowledge that being a savvy business operator is more important than being a skilled tradesperson.

In his bestselling book *Selling to Serve*,[14] James Ashford emphasises that the primary function of any business is to make a profit. Although profit might not have been the initial motivation for starting a business, it remains the core objective of every business endeavour. Profit is derived from the difference between revenue and expenses. If a business fails to generate revenue that surpasses its expenses, it will eventually exhaust its cash reserves and face financial difficulties.

The path to a successful business is paved with profit, providing financial stability and avoiding the hardships associated with unprofitable enterprises. Therefore, it is crucial to reframe our perception of profit and recognise it as the primary function and driving force behind every successful business.

Fear of fat cats

The expression 'fat cats' originated in the United States, where it was first used in the 1920s to describe wealthy, powerful individuals and political donors. The term carries corrupt connotations, evoking images of greed and avarice. Characters like Gordon Gecko from the *Wall Street* movies, the 'Wolf of Wall Street' Jordan Belfort, Montgomery Burns from *The Simpsons* and Walt Disney's Scrooge McDuck embody this stereotype, portraying individuals who prioritise money over human relationships.

However, it is important to distinguish between making a profit in business and losing one's moral compass. While nobody wants to be seen as a heartless fat cat, equating profitability with a lack of ethics is a misconception.

CASE STUDY: Geoff and Lyn make some tough decisions

Geoff and Lyn were busy running their powder-coating business, and for some years had been earning less than their employees, leaving little profit in the business. Both worked long hours, but felt ridiculous working so hard given their age.

When they came to Next Level, we analysed their business numbers and formulated a strategy within the budget for 25/10: pay them both a commercial wage while the business makes a 10% operating profit. To achieve this, they needed to make tough decisions, such as letting low-margin customers go and raising prices. However, by involving Geoff and Lyn in the budgeting process, my team gave them confidence to embrace these tough decisions.

Today, Geoff is paid a $130,000 salary for working full-time in the business, while Lyn is paid $65,000 for working part-time. All that and the business makes a 14% operating profit – a lot better than the $80,000, $30,000 and 2% they made before.

Understand your numbers to make a profit

Good trades businesses continually refine their offerings to deliver exceptional service. They understand the importance of ongoing improvement to meet the evolving needs and expectations of their clientele. However, it is equally important to pay close attention to improving profit margins. This includes regularly reviewing charge-out rates, pricing methods, markup rates and payment terms.

If your business's bookkeeping and accounting function comes up short, you will be unable to measure the financial impacts of providing higher levels of service and quality to customers. Consequently, although your business is improving its value offering, the price it's charging falls behind the value provided.

I've seen businesses at serious financial risk, despite delivering services at a higher standard than the competition. I've seen dwindling cash reserves and profits, even though demand was soaring. Why? All because value exceeded the price charged.

Businesses in this situation are unable to grow and suffer from financial strain. Without profit, the business loses its value. It can be disheartening for owners to invest so much time and effort, only to find their business holds little worth.

Knowing your numbers is essential for making a profit and ensuring sustainable growth and success in business. By closely monitoring and analysing financial figures, you can make timely decisions, identify areas for improvement and take necessary actions to increase profitability. Operating profit (10% plus) must be a non-negotiable.

Business growth requires profit

Profit is a growth enabler. It enables you to provide customers with the highest levels of service. It enables you to invest in the machines you need to complete higher-quality work in less time. It enables you to hire a project manager, an operations manager or chief executive officer (CEO). Without profit, all of the above is impossible.

As we saw in Chapter 2, profit is unselfish – many rely on your business to succeed. Profitable businesses pay bills and meet financial obligations. The Reserve Bank of Australia noted only 60% of small businesses in the country were profitable in the financial year 2009/2010. Don't be a business in that other 40%.[15]

The profit trap

As we've discussed, excelling at your work while running an unprofitable business is caused by a value/price misalignment. If that's your situation, you may be in a profit trap.

You can get stuck in a profit trap when your customers are highly satisfied with the service you provide at the current price. They not only recommend you to others, but also express disbelief at how you can offer such an amazing service at such affordable rates.

Now the trap is set. Referrals pour in, but these referrals have been pre-conditioned to expect low prices and high value, just like their friends received. If you propose a fee structure that properly aligns value and price, they reject it because it doesn't meet their low-price expectation.

You are trapped. Increasing prices to match value is daunting. You fear customers will not accept the price hike. Your limiting beliefs perpetuate the profit trap, making it even more difficult to raise prices. You tell yourself all sorts of reasons why it can't be done.

The profit trap is especially disastrous for construction businesses dealing with project work. As projects often run over six-month to eighteen-month time horizons, setting the wrong price will come back to beat you over the head, again and again.

You can see here how understanding numbers is essential for effective pricing, appropriate profitability and avoiding the profit trap.

Discounts

A study of over 6,000 companies by ProfitWell found those that offered customers a discount had more than twice the customer churn rate of those that didn't.[16] Furthermore, the churn rate increased in proportion to the discount level. The study also found only 70% of customers were willing to pay the full price after previously being provided with a discount.

This study suggests providing customers with discounts erodes product or service value because cost-conscious customers will likely churn. It also suggests that seven out of ten customers will accept paying full price even after being provided a discount. Keep this in mind as we further explore why you should think twice before offering discounts.

CASE STUDY: Peter and Brett's discounting conundrum

Peter and Brett spent two days preparing a quote for a nine-month plumbing contract, which amounted to $750,000. When they presented the quote, the client expressed interest in hiring them, but mentioned that another plumbing contractor had offered a price $37,500 lower. The client proposed that if Peter and Brett matched the competitor's price, they would be awarded the job.

Peter and Brett had two options:

- Accept the client's offer and match the competitor's price. Despite the profit being reduced by $37,500, the contract would still amount to a decent $712,500 in revenue over a nine-month period.
- Decline the client's offer to match the competitor's price. A $37,500 discount reduces profit and requires additional work to make up for the shortfall.

What would you do? A or B? Based on my experience, 80% of trade and construction business owners would choose option A, assuming that $37,500 is only a small reduction off the $750,000 contract value.

For all of you that chose B, you are correct. Here's why.

Peter and Brett's business ran on a 10% operating profit margin, meaning $75,000 on a $750,000 job. If they discounted the contract by $37,500, they'd have been halving their operating profit from 10% to 5% ($75,000 reduced to $37,500). Should they agree to the discount, they'd have to do two $750,000 jobs at that discount, instead of one.

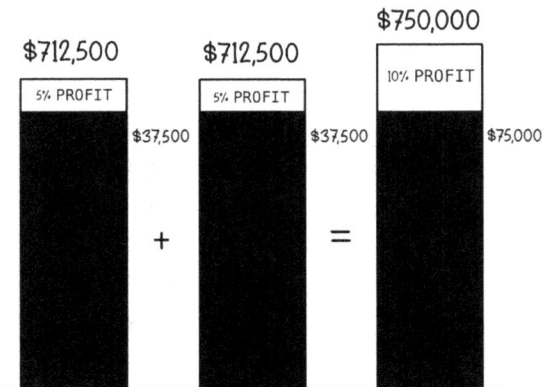

Peter and Brett's discount conundrum

What did the boys do? Peter and Brett declined to match the competitor's discount and didn't budge on their price. They understood their value and were unwilling to negotiate on price, even if it meant losing the contract.

The good news is, the boys won the plumbing contract on their price terms.

PRO TIP

Discounting is the act of providing value that exceeds the amount charged. However, it is important to note that discounting is not a viable financial strategy. In fact, it can lead to disaster. With discounts come higher workloads and smaller margins.

The case study of Peter and Brett demonstrates the importance of understanding your numbers and the profit margins required to achieve your business's financial goals. If the best you've got is gut feel – good luck.

To ensure the financial success of your business, it is crucial to protect margins. Take a moment to reflect on your business's capacity and consider how many 5% profit jobs it can handle in a year. Time is precious, and you cannot afford to waste twelve months. It is imperative to thoroughly evaluate customers, projects and associated profit margins. By doing so, you can identify and pitch for the projects that will contribute to your business's sustained profitability.

There's nothing wrong with a customer asking for a discount – in fact, good on them. When a client asks for a discount, you may think they are saying your price is too high, but they aren't. They've just asked for a discount.

Price is only one consideration in the buyer's decision-making process. How many times have you found yourself purchasing items such as a new washing machine, car, TV or piece of jewellery, only to realise that you ended up paying more than your intended budget? Although the buyer may not want to pay your price, they may end up paying it anyway because they recognise the value of your service.

When you hold firm on price, you convey the belief that you know what you are doing and understand the value your business provides. Holding firm on price is holding firm on your margins, which in turn allows you to bank profits, make further investments in your business and provide higher levels of service to your customers. Higher profit margins enable you to increase customer satisfaction, which attracts more high-quality clients.

Money and mental health

A 2022 report by the Australian social research consultancy firm, Heartward Strategic, found a significant link between money and mental health.[17] People who are experiencing financial difficulties are twice as likely to be experiencing mental health challenges too.

Another report by McNair yellowSquares found that 34% of small business owners in Australia received a diagnosis of either stress, depression or anxiety in 2019.[18] The main reasons found to be behind this were concerns about:

- Profitability and survival of their business

- Maintaining cash flow

- Attracting and retaining customers

- Potential impacts on family

These two statistics suggest that financial wellbeing plays a crucial role in overall mental wellbeing. Having a profitable business with decent margins and cash flow is good for your business finances *and* it makes you happier and less prone to mental health issues.

Cutting corners to make a profit?

Imagine you're the CEO of a demolition business. Your operations manager presents a proposal to you. They suggest lowering the quality of work, reducing safe workplace practices, cutting employee pay, using deceptive marketing and ignoring environmentally safe waste disposal, all in the name of saving money. Would these changes lead to sustainable business profits?

Hopefully, you answered no. Poor business profitability may lead to similar corner-cutting cost choices in your business.

Let me explain. Without profit, you are forced to compromise on quality by using inferior materials, cutting corners on safety, underpaying your staff, sending out misleading marketing messages and disposing of waste in an environmentally unsafe way. Without profit, you'll lack funds and the best you will be able to do is to tread water, until your customers churn, your employees leave and warranty claims arise. Soon, your reputation could be irreparably tarnished, which will likely be disastrous for your mental health.

Never be a corner cutter.

Summary

Profitability is the primary function of any business. Failing to address profitability issues can result in serious problems, such as cash-flow difficulties, high levels of debt and insufficient working capital. Left unaddressed, these issues can ultimately threaten the survival of your business.

Neglecting to invest in the accounting and finance function of your business means you'll lack knowledge with important business numbers. This could lead you to prioritise revenue over profit. It is vital to gain comprehension of your numbers to improve profitability.

In Part Two, I will demonstrate an easy way to reverse the tide of business profitability. This involves adopting a clear and logical framework for assessing and addressing your business's financial issues. First, though, let's look at exactly why you need a budget for your business.

FOUR
Why You Need A Budget

In our journey so far, we've delved into the five biggest mistakes trades and construction businesses make with their finances. In Chapter 2, we explored the profound significance of your business, not just for you, but also for your family and the community. Chapter 3 emphasised the undeniable importance of profit as the true indicator of your business's vitality. Let's continue this exploration together and discuss why you need a budget.

There are two non-negotiable promises of this book to deliver 25/10. You will:

- Pay yourself $2,500 per week

- Design your business to make a 10%-plus operating profit

How do you do that? With a budget.

The budget for your business will be a game plan. It will serve as your blueprint to achieve financial success. It will act as a model upon which you can build. Most importantly, it will protect your business's financial health. In short, the budget is the missing rungs of the ladder you climb to unlock your true business potential.

Climbing the budget rungs of the business ladder

What is a budget?

In business terms, a budget is a financial plan that outlines the projected income and expenses over a specific period. It serves as a tool to anticipate and plan for expenses, allocate resources, set financial goals, track performance and make informed decisions to ensure the financial health and continuity of the business.

A well-structured budget is fundamental to your finance and accounting function, showing the way to navigate the financial landscape of your business. It enables you to monitor and gain a clearer understanding of whether your business generates sufficient income to cover its expenses, and to take necessary actions. Without a budget, you are essentially operating in the dark, risking financial instability and hindering growth and success.

The right tool for the job

The right tools are essential in your business. A wrench, bandsaw, impact drill, tipper truck all carry out a particular function on your jobs. Tradies love their tools and ensure they always use the right one.

Tools increase efficiency. For example, an air gun will nail fence palings maybe ten times faster than a hammer. However, onsite tools are distinct from offsite tools. In the fast-paced world of trades businesses, offsite tools are often neglected. Unfortunately, this leads to oversight of the most influential tool of all – the budget.

A budget shows you how to achieve your financial goals. It is a tool that helps you to identify problems early, enabling you, the boss, to act sooner rather than later or, worse, too late.

Your financial scales

Having a business budget is like using scales to control your weight. Just as scales provide you with a clear measurement of your progress towards weight loss, a business budget provides you with a quantifiable measurement of your financial progress.

With scales, it's simple to know whether you're gaining or losing weight. That helps you make informed decisions about your diet and exercise routine. Similarly, when you refer to a business budget, you can monitor your progress and define areas where you need to cut back or invest more. You get high-quality information to make the best decisions about allocating resources and managing expenses.

You may have lapses – that piece of chocolate cake yesterday was irresistible, and that time spent at the coffee shop meant you skipped a gym session. Just as the scales tell all, so it is with a budget, which reminds you to stay within financial limits.

With a budget, you will rise.

Top ten benefits of having a business budget

1. **Strategic guidance**. A budget as a blueprint for your business provides a clear and strategic path to achieve your goals. It guides you in protecting your business's financial health.

2. **Resource allocation**. A budget allows you to allocate resources effectively. It helps you anticipate and plan for expenses, ensuring that you have the necessary funds available when needed.

3. **Financial goal setting**. A well-crafted budget helps you set financial goals. It provides a framework to track your progress towards them.

4. **Performance tracking**. By comparing your actual income and expenses to those budgeted, you can follow the business's financial performance, enabling you to identify areas for improvement or potential issues early.

5. **Informed decision making**. A budget provides you with the figures to make informed decisions. It helps you evaluate the implications of different choices and prioritise investments or cost-cutting measures accordingly.

6. **Financial stability**. Operating without a budget puts your business at risk of financial instability. A well-planned budget ensures that

income is sufficient to cover expenses, reducing the likelihood of cash-flow problems.

7. **Continuity and growth**. A budget plays a crucial role in ensuring the continuity of operations. It helps you identify potential risks, plan for contingencies and make adjustments to ensure long-term success and growth.

8. **Accountability**. A budget keeps you accountable for your financial decisions. It serves as a tool to measure the financial implications of your choices.

9. **Empowerment**. Embracing a budget empowers you to take control of your business's financial health. It gives you the tools and information needed to proactively manage finances and make the right strategic decisions.

10. **Opportunity for innovation**. A budget encourages creativity and innovation. It challenges you to find cost-effective alternatives and discover new ways to achieve your objectives without compromising quality.

Overall, having a well-crafted business budget is essential for managing finances effectively, ensuring stability and unlocking the true potential of your business. In any business, a budget is a must.

CASE STUDY: Geoff and Lyn's budget

Geoff and Lyn had been in business for two decades when they came to the team at Next Level. They were concerned about their numbers and knew they lacked a financial plan that would enable their business to generate higher income. We immediately sought to address their concerns by analysing their transactions, margins and financial results.

It didn't take long to discover they had problems with margins and resource allocation. The margins were too low, and the site work division was pulling resources (labour and cash) away from the factory division.

Our recommendation? A budget.

We collaborated with them to create a detailed 25/10 budget based on their projected work and anticipated costs. While the concept seemed simple in theory, the actual process proved to be technical. Geoff and Lyn, who had no prior experience in budgeting, were challenged to think in new ways. To their credit, their desire to succeed was strong.

Initially, they experienced a mix of apprehension and anticipation, as is common with any new undertaking. However, as the months progressed, and the figures consistently met targets, excitement and relief took over. Geoff and Lyn were confident that both their income and expenses were on track. As a result, they exceeded expectations in the first year.

Geoff, Lyn and the entire team at Next Level were stoked.

The seven steps to preparing your budget

Step 1: Thorough review of records. Review your business's financial records, including revenues, margins, overhead expenses and direct costs. This provides insights into the financial health and performance of your business.

Step 2: Project sales. Check prior year sales, market trends and your industry knowledge. From that, forecast anticipated sales for the next twelve months. This helps you understand revenue streams and growth opportunities.

Step 3: Work out your GPM. Work out the GPM for your business for the next twelve months. To do this, check prior year GPM and use your judgement when setting a GPM for the budget.

When the GPM is determined for the budget, calculate gross profit by multiplying total revenue by the GPM.

Step 4: Allocate your fixed costs. Once you have a clear understanding of your fixed costs, allocate these expenses to the relevant expense account on the budget.

Step 5: Set clear and specific goals. Define and write down your financial goals. They may be to pay yourself a wage and allocate a 10% operating profit back into the business. They may also be to improve GPM and the operating profit percentage. Have a time-frame; twelve months is usual.

Step 6: Calculate and assess your profit and loss. Calculate your operating profit by subtracting total operating expenses from total revenue. This assesses your business's financial viability and is key to determining whether an operating profit of 10% or more can be achieved.

Step 7: Monitor, review and adjust. Track actual income and expenses against budgeted amounts. Ongoing monitoring allows you to identify deviations and make necessary adjustments. The bonus is you become flexible and adaptable while maintaining financial stability and success.

Remember, the creation and regular tracking of your budget to actual performance is essential. A VCFO will provide the detailed intel, and so armed, you can make informed decisions and take proactive measures to ensure the sustainability of your business.

Once the budget is in place, it becomes the money plan for your business, allowing you to gain deep understanding of your key financial numbers. With the budget as a reference, you can confidently assess your performance and align your priorities accordingly. It empowers you to make informed decisions, both onsite and in the office, to better steer your business towards its targets.

That's when you are getting the full benefit of a VCFO service.

Virtual chief financial officer

Many owners of trades businesses accelerate their financial results by enlisting the services of a professional VCFO. Your VCFO will hold your business performance accountable to the budget – much like a project manager on a construction site would hold the production schedule to a standard, the VCFO does it with numbers. If the numbers are not reached, they'll investigate the reasons why.

A VCFO is most impactful in the early stages of a business's financial restoration where you are overwhelmed and need all the help you can get. They will analyse your performance against target and direct you to areas where you may be losing money. In time, you will see where they are coming from, and so gain a good understanding of the numbers yourself. That is when you will excel.

CASE STUDY: Mick's VCFO

Mick's construction company had a revenue shortfall of $75,000 per month. His VCFO examined the numbers to pinpoint the cause of the problem.

The business had a revenue target of $300,000 per month – realistic based on past performance, new customer enquiries and market trends. However, three months into the financial year, sales revenue was only achieving 75% of forecast. Instead of $300,000 per month, Mick's business was only making $225,000.

His first thought was to ramp up business and win more projects, but he already had contracts lined up for the next eighteen months. What should he do? He set his VCFO to work.

Through analysis of numbers and production schedules, the VCFO determined the problem not to be sales, but rather production efficiency. The business was geared to do no more than four jobs simultaneously, but Mick had five on the go. In other words, there was a 20% increase in production beyond capacity. As a result, pressure mounted on scheduling and production.

That caused a chain reaction. The schedule of onsite tradespeople was messed up. The plasterer arrived before the concreter had finished, and the concreter started late because the plumber hadn't laid the pipes on time. As a result, stages of completion were delayed and, consequently, the dates on which invoices could be raised were affected.

The VCFO service allowed Mick to step back and objectively review his predicament. Upon digesting all the information, he traced the problem to its source – himself. He caused the scheduling logjam by taking on more projects than he could deliver.

What was he going to do about it? To arrest the immediate problem, Mick needed to halt production on one project, thereby allowing him to speed up overall production and increase cash flow on the other four.

Halting production on one project with six more in the pipeline meant work was piling up. It became essential to hire a project manager and expand production capacity. This required budget adjustment.

It took a few months, but he found the right person. His new hire took over management responsibilities for six of the outstanding projects, reducing Mick's workload to just one. This freed up Mick's time to focus on the business, which he did. He took the opportunity to fix production procedures he knew needed fixing, but had never had the time to address. In the process, he learned more about time management, the numbers and himself.

The result? Measurable improvement. Production capacity increased. Work finished ahead of time. Stage of completion invoices went out sooner. Money came in earlier. However, the most significant improvement was the reduction in overall confusion, disorder and stress, which was immeasurable.

Mick had arrived.

Three budget levers

Mick's experience highlights the importance of a business budget, so it is alarming that eight out of ten business owners operate without one.[19] Without a budget, these businesses blindly navigate the next twelve months, relying solely on hard work to earn a fair wage and hopefully generate profits.

This is where an investment in a VCFO service delivers financial rewards. The VCFO provides expert guidance and strategies to help businesses establish and manage budgets, optimise their financial performance and make informed decisions for long-term success.

As mentioned in Chapter 1, research indicates that only one in five small business owners considers themselves to be a good financial manager.[20] The ones that see themselves as good financial managers all have a well-documented budget. These entrepreneurs understand the three key levers that can be exploited to their financial advantage:

- **Revenue.** Determine the monthly revenue required to meet your financial goals. Think about your production capacity / limits, projected production schedules, production efficiency, availability of trades, price and seasonality to name a few considerations. Adjust revenue numbers to factor in these considerations.

- **GPM.** Experimenting with the GPM, you can begin to get an idea of the right combination of that and revenue needed to achieve your financial goals. This exercise will give you valuable insights into the correlation between revenue, GPM and operating profit.

- **Overhead expenses.** There will be instances where increasing your overhead expenses is necessary. For example, if you're expanding your business, you may need to make a provision for wages and additional rent for extra factory space.

Review your overheads and identify unnecessary expenses. For example, consider cancelling subscriptions you no longer need and reassessing your gas, electricity and insurance suppliers, especially if they haven't been reviewed in years.

Pulling the levers

Combining these three levers allows you to evaluate, plan, adjust, re-evaluate and run scenarios. For example, you can simulate the effects of increasing your GPM from 50% to 52%. You can also examine the implications of a 5% decrease in revenue. Get a feel for which levers need adjustment to streamline operations, improve efficiency, handle emergencies and ultimately generate higher profits. Stay committed; rewards are only as consistent as the effort invested.

PRO TIP

Many business owners create budgets but fail to review them because they remain buried as a file on a computer. Get yours professionally printed in size A0 or A1 and gloss laminated. Take ownership. Display it on your office wall, easily accessible without you having to search through your computer. Let your staff know the budget target and actual outputs, daily, weekly or monthly – whatever basis you decide. With your team onboard as you lead the way, your business is ready to go places.

Ninety-day action plans reviewed quarterly

Once your budget is in place, set a ninety-day action plan and highlight the goals it aspires to. Those businesses with twelve-month financial goals often aim to:

- Grow revenue

- Increase GPM

- Increase operating profit

- Optimise expenses to match financial plans

CASE STUDY: Peter and Brett's ninety-day plan

Peter and Brett suffered frustration due to the lack of a financial plan showing the path to profitable growth. When they came to the team at Next Level, we set a twelve-month budget plan with a particular focus on ninety-day financial growth.

Achieving this goal required Peter and Brett's business to:

- Grow team size from eight to ten

- Become a major sponsor of the community football club

- Develop a pricing methodology to lift hourly charge-out rates by 10%

- Increase markup on materials from 20% to 25%

- Implement an administration charge on every contract variation

Three months into the financial plan, when I met with the boys in our VCFO session, I asked them how the ninety-day action plan was going. Brett reported all ninety-day action items had been implemented and they were already starting to see positive results. With team expansion, they were able to take on more projects and increase revenue.

Furthermore, he said, 'Becoming a major sponsor of the community football club helped raise brand visibility and attract several project enquiries. The pricing methodology and markup adjustments allowed us to increase our profitability, while the implementation of an administration charge helped cover the time involved in contract variations.'

It was plain to see that overall, the financial plan and ninety-day action plan were proving to be effective in driving profitable growth for their business.

How does a VCFO service work?

At Next Level, the team's process works as follows.

Setting the 25/10 budget

First and foremost, a business needs a budget. It's important that a business owner understands their budget and has input into assumptions and forecasts. It must be realistic and achievable.

When the numbers are in the range, we look at the GPM, revenue and overheads, and adjust those three levers until the owner hits the target of a $2,500 per week wage and 10%-plus operating profit (25/10). We then set a ninety-day action plan to capture the short-term goals for the business, just like Peter and Brett who increased materials markup from 20% to 25% and employed two extra staff. We finalise budget, and with everyone onboard, the business owner gets it up on the wall.

Monthly meetings

A month into the ninety-day plan, we prepare end-of-month accounts and financial reports. Examining the numbers, we call a meeting with the business owner and discuss results. It's exciting.

The Next Level method reveals key financial indicators and their effect on financial performance. We highlight both the good and bad, including notable wins or losses that might have caused them. We review the short-term goals on the ninety-day action plan and assess how the business is tracking. The whole process is a much-needed nudge of accountability to those important business goals that can get lost in the day-to-day.

Looking ahead, we consider the external situation. This includes upcoming economic news such as softening demand, increasing interest rates, changing material prices or energy price increases. It is important to reflect on how these developments may impact the coming months, allowing the business time to establish contingency plans.

To achieve the best results, we recommend monthly meetings for business owners to stay on top of all variables in the fast-moving economic climate. This ensures finances stay fresh in mind, enabling impactful decisions.

Tax time

In April, we engage in tax planning, which includes creating a forecast and recommending options to legally reduce taxes. This process involves reviewing the previous budget and its outcomes, considering current performance and end objectives, and resetting the budget for the next twelve months. Despite sounding complex, this becomes clearer when we put it in figures.

Results

This is the key issue.

Results may not be apparent in the first three months, but over time, noticeable trends emerge. With continuous improvements and adjustments, financial performance begins to take off.

We love hearing from our clients how they are gaining traction in their business. They'll say, 'This is the first contract I have won with a 25% markup' or 'This is the first deal I've made with an administration charge clause written into the contract.' That's music to all ears.

When the whole team in a business is committed and responsible for executing the plan, improvement is inevitable. By staying dedicated for a two-year period, you can ensure your business achieves remarkable progress and there will be no regrets looking back.

Summary

My greatest hope is that you are now convinced of the necessity of a budget. Adopt one and let it be the key pillar of your business financial plan. However, to truly amplify your progress and achieve remarkable results, hire a VCFO. With their expertise, proven methods and track record of helping businesses similar to yours, a VCFO will provide invaluable financial guidance, insight and support.

These recommendations will help you navigate any economic climate and keep you financially aware of your numbers. VCFO services are designed to fulfil one promise: 25/10.

PART TWO
THE CLEAR METHOD

FIVE

Clean-up

The CLEAR method is a tried and tested five-step process, developed and honed by Next Level over twenty years. The purpose of the CLEAR method is to transform your business into a well-oiled, high-performance profit-making machine.

CLEAR is an acronym for the five stages every business owner needs to pass through to gain full control over finances – a journey from chaos to financial clarity, hence the term CLEAR. These stages are:

- **C**lean up the financial chaos
- **L**earn the financial processes
- **E**njoy the fruits of your efforts

Clean up
Identify and clean up the issues causing murky financial visibility

Learn
Set-up the finance and accounting functions for financial visibility

Enjoy
Confidence in knowing your numbers and positive cashflow, enjoy!

Ambition
Chase your dream and reset goals, backed by sound financial management

Reward
Legacy, transition, succession – your choice, your reward!

3–6 Months

6–12 Months

12 Months

1–4 Years

∞ Infinity

Making money

High

Financial visibility

Low

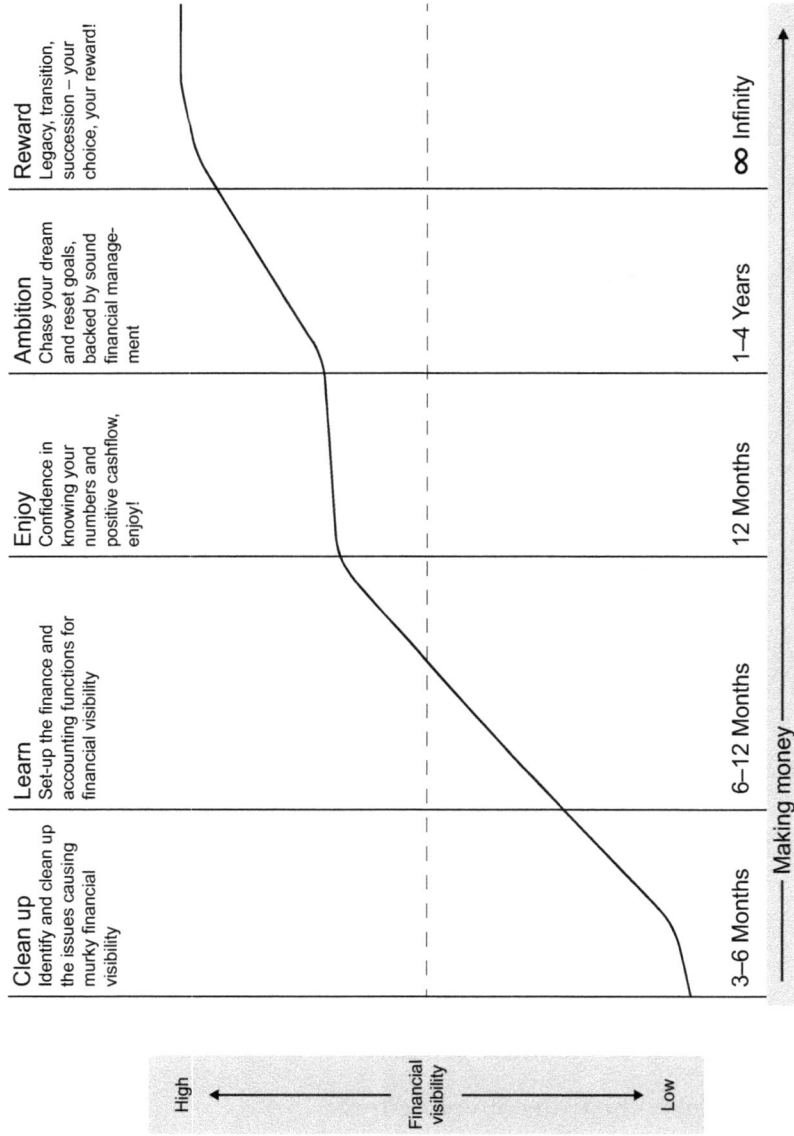

The stages of the CLEAR method

- **A**mbition to take it to the next level

- **R**eward yourself as stewardship, legacy and succession loom

Visit https://nextlevelaccountants.scoreapp.com and complete the Next Level Scorecard to identify which stage your business is at.

If your business is in the Clean-up phase, the CLEAR method will expedite progress and help you reach the Learn phase as rapidly as possible. This chapter provides an inventory of the necessary tasks to address before your business can advance.

Completing this phase, although not enjoyable, is essential under the CLEAR method to achieve financial visibility and increased earnings.

Are you in the Clean-up phase?

Businesses in Clean-up are in a dire financial mess. Financial management is a constant struggle, with each day bringing new challenges. There is a multitude of unending money troubles, which serve as painful reminders of the deep-rooted financial issues plaguing the business.

You are in the Clean-up phase if you:

- Are not paying yourself $2,500 per week

- Have no idea what your GPM is

- Don't know what your operating profit is

- Are struggling with cash flow

- Are not sure how to read financial reports

- Are behind in superannuation entitlements

- Do not track employee entitlements, such as annual and sick leave

- Manage bookkeeping as an afterthought

- Are overdue with business activity statements (BAS)

- Are overdue with income-tax obligations

- Don't know / have a budget to forecast profit

Trapped in the Clean-up phase

A 2015 poll of small business owners in the United States found that 40% mentioned bookkeeping and taxes as the worst part of owning a business.[21] A similar percent said they spent more than eighty hours a year – two weeks full-time work – on bookkeeping and tax preparation tasks. Clearly, many business owners are frustrated at managing tasks that are not in their skill set, unaware they may be creating more problems than they are solving.

Why do they do this? Many business owners mistakenly view employing an expert for bookkeeping, tax preparation and financial management as an expense, rather than an investment. To keep expenses low, some will do it themselves, while others rely on their spouse or entrust their retired father who's supposed to be skilled with numbers. Even worse, they turn to a well-meaning but inexperienced friend of a friend. These choices trap businesses in an unending cycle of financial stress and confusion.

The best way out, maybe the only way out, is to get professional assistance.

The four stages of the Clean-up phase

Imagine you've called the team at Next Level. What will we do? We'll guide you through the four-part process of the Clean-up phase, the C of CLEAR:

1. **Tax-office compliance** – facing the music

2. **Bookkeeping makeover** – cleaning up the clutter

3. **Addressing overdue tax lodgements** – redirecting from chaos to compliance

4. **Tax payment plans and strategic solutions** – conquering confusion and reclaiming control

Let's have a look at each stage in detail.

Step 1: Tax-office compliance

In the first step of the Clean-up phase, we conduct a tax-office compliance audit to assess your tax obligations, including all overdue tax lodgements. This involves contacting the tax office to determine the date of the last lodgement for all required tax forms and any that are overdue. This assessment ensures nothing is missing in fixing your tax compliance.

These tax forms include:

- Company income-tax returns
- Personal income-tax returns
- BAS
- Instalment activity statements (IAS)
- Single touch payroll (STP) finalisation
- Taxable payments annual report (TPAR)

Tax terms

If that list was confusing to you, here's a quick refresher on the meaning of some of these tax terms. These terms apply in the Australian context, although many countries have similar reporting requirements:

- **BAS**. This is a form that Australian businesses submit to the Australian Tax Office (ATO) to report and pay their GST and pay as you go (PAYG) withholding obligations. The BAS is typically lodged quarterly. BAS includes wine equalisation tax, fuel tax credits and fringe benefits tax, if applicable.

- **GST**. In Australia, the federal government imposes a 10% consumption tax, the GST. When you sell goods or services for $50, the customer pays $55, with the additional $5 being the GST. Similarly, when your business buys supplies, you pay 10% GST, which can be claimed back as a credit. When completing a BAS form, you report the GST amount collected on sales minus the GST paid on purchases to determine your refund or amount you owe to the tax office.

- **PAYG withholding**. Businesses are required to withhold a certain amount of tax from their employees' wages or salaries and remit the withheld amount to the ATO on the employees' behalf. The withheld amount is based on each employee's income, tax file number declaration and other relevant factors. This system ensures that individual employees meet their tax obligations throughout the year rather than paying a lump sum at the end.

- **IAS**. This form covers the reporting and paying of PAYG withholding on non-BAS months.

- **STP finalisation**. Australian employers must report on employees' salaries / wages, super contributions and PAYG withholding amounts to the tax office with each pay run. With STP-enabled payroll or accounting software, employers can submit information to the tax office automatically each time they process their payroll. At the end of the financial year, they collate and submit the actual figures, which is called STP finalisation.

- **TPAR**. Businesses in trades, construction and several other industries that make payments to contractors need to report these payments and lodge a TPAR every year.

- **Superannuation guarantee (SG) charge**. If you fail to pay employees' entitled SG to the correct super fund by the due date, you are required to submit a SG charge statement to the tax office for each quarter in which the payment was late or overdue. Additionally, you must pay the SG charge, which includes the overdue superannuation amount, an interest amount applied to that superannuation and an administration charge.

Get a pro

Unless you're a trained accountant, all this can be overwhelming. That's why it's essential to seek professional assistance. The stress experienced by business owners is often paralysing, leaving them feeling stuck and helpless. Many try to ignore the problem, hoping it will magically disappear, but the truth is, it never does. The only way to break free from these obligations is to summon the courage to confront them head-on and address the situation, even if it's difficult and uncomfortable. A pro can ease your burden.

Step 2: Bookkeeping makeover

After establishing your overdue tax compliance obligations, we can move on to step two: cleaning up the bookkeeping. This involves fixing the business's bookkeeping to ensure accurate recording, categorisation and reconciliation. If discrepancies are found, and they usually are, they need to be adjusted and corrected.

The problems often lie in these areas:

- Debtors' ledger (sales)
- Creditors' ledger (purchases)
- Payroll
- Bank reconciliations

The debtors' ledger

This provides a detailed list of all unpaid customer invoices (aged receivables). The bookkeeping needs to be cleaned up to ensure that when the aged receivables report is run in your accounting software, each unpaid invoice displayed is accurate. It is important to determine which customers owe you money and how much of that amount is overdue. Never forget, incoming cash is king.

The creditors' ledger

This provides a detailed list of all unpaid bills – it's what you owe to your suppliers. An important element of this report is in ensuring your suppliers agree to what you say is owed.

Payroll

Your business must ensure that it pays employees according to their contractual or award terms and conditions, and that records of any owed entitlements are accurate. For instance, both part-time and full-time employees accumulate annual and sick leave. You must have the correct employee entitlement balances to avoid overpaying or underpaying entitlements.

Payroll has several subsections:

- **Employee details** must be kept up to date, including personal details, names, addresses and emergency contacts.

- **Employment settings** must be current. This covers whether each employee is correctly classified as full-time, part-time or casual, the date they were first employed, and their payroll calendar and superannuation member account details.

- **Tax settings**. Is each employee an Australian resident or a foreign resident for tax purposes? Did they elect to claim the tax-free threshold? Do they have a student supplement support loan or a higher education contribution scheme debt? Ensure tax settings reflect all disclosures.

- **Leave**. Entitlements accrue in line with the number of hours worked and are sometimes misaligned. For example, if an employee is working a thirty-eight-hour week, ensure their annual leave isn't accruing based on a forty-hour week – a small but costly detail.

- **Superannuation payable**. Reconcile this amount to each employee to avoid over or underpayment. If it is overstated, you are paying more superannuation than necessary. If understated, you may need to submit an SG charge statement.

Bank reconciliations

A bank reconciliation is a process of comparing the balance in an organisation's bank statement with that in its own accounting records. It involves reviewing and matching each transaction recorded in the bank statement with the corresponding transaction in the company's books.

The purpose of a bank reconciliation is to identify discrepancies or errors between the two balances and to ensure that the company's financial records accurately reflect its actual bank transactions. This process helps detect and resolve any missing transactions, bank errors or unauthorised activities. By performing weekly bank reconciliations, business owners can maintain accurate financial records and have pinpoint understanding of their true cash position.

CASE STUDY: Mick's bookkeeping makeover

Mick was working long hours in his construction business and his partner agreed to do the books, to help him out and reduce his workload. The problem was she wasn't a bookkeeper, she didn't like bookkeeping and had a lot on her own plate, including working a part-time job and being a mum. The bookkeeping always took a backseat to other priorities in her life.

When the Next Level team came in and reviewed the bookkeeping, we identified numerous problems with the debtors, creditors and bank accounts, which resulted in an inaccurate picture of Mick's true business finances. Once Mick handed it to us, we were able to fix the errors and provide him with the true picture.

Step 3: Addressing overdue tax lodgements

With debtors, creditors, payroll, superannuation and bank accounts reconciled, you can move on to step three, which is to lodge those overdue tax documents identified in step one.

Your tax return

A tax return is a form or document that individuals, businesses and organisations file with the ATO to report their income, expenses and other relevant financial information. It is used to calculate the amount of tax owed. All tax payers must file a tax return annually to comply with the law.

> **PRO TIP**
>
> If you do your tax return yourself, in Australia, you have four months to lodge it – your deadline is 31 October. If you appoint a tax agent, you get an extension until 15 May the next year, which gives you an extra six and a half months to get your affairs in order (and gather the money to pay any additional tax).

Lodgement options

Australian taxpayers have three ways to lodge a tax return:

- On paper
- Online through MyTax
- Through a tax agent

In addition to the tax lodgement due-date extension, a tax agent provides you with professional advice, which will reassure you that all eligible income and deductions are being claimed accurately. Moreover, your risk of a random tax audit is reduced – a huge weight off the shoulders.

It's worth noting that in Australia, only registered tax agents can charge a fee for the preparation and lodgement of tax returns. Many professional bookkeepers are not tax agents.

CASE STUDY: Peter and Brett's accountant upgrade

As Peter and Brett grew in their business, they outgrew their accountant – a mate. While reminding them of their overdue lodgements, he didn't push the issue. No one was happy.

Peter and Brett called the team at Next Level. We identified they were in the Clean-up phase of CLEAR and, upon getting the go-ahead, jumped on to their books, accounts and tax-office lodgements. With that sorted out, they were ready for the next four phases – learn, enjoy, ambition and reward, which as you will see can be thrilling.

The message here is to bite the bullet and move forward.

Facing the tax debt monster

Preparing and lodging all outstanding tax returns will reveal where you stand in terms of your tax debt obligations. For many business owners, that can mean a tax bill with a great big number on it.

In my experience, this is the moment business owners dread, but there's no getting around it. At some point, the tax office sends nasty letters demanding lodgement and threatening legal action. Then the nightmare begins.

Far too often, unresolved tax debts become the downfall of businesses, leading to failure and insolvency. Don't let that be you. Take action now. Face the tax debts before they spiral out of control and the axe descends. Your business and financial stability depend on it.

Step 4: Tax payment plans and strategic solutions

With tax lodgements complete, magic happens. Tax agents, with their expertise, step in to assist you in effectively managing your tax debt. No longer do you need to feel overwhelmed or crippled by fear. With their guidance and support, you can confidently tackle these challenges within your business and regain control.

Tax agents will negotiate with the tax office on your behalf. They explain the reasons for your non-compliance – perhaps a combination of previous bad advice, underinvestment in bookkeeping and / or health issues. If you've been historically behind on tax lodgements, engaging a tax agent to get your book-keeping cleaned up is a statement signalling you'll meet obligations and play by the rules from now on. This is a game changer.

After all, no one – you, the tax office or the economy – gains by your non-compliance. If you are willing to play by the rules, the tax office is willing to negotiate.

Tax payment plan

Upon you committing to debt settlement, the ATO may accept a payment plan proposal, which allows you to gradually pay off your debt in instalments over a period not exceeding twenty-four months. Having the plan agreed enables you to get on and run your business without the fear of being crippled by a large one-off tax debt. That's a huge positive for you.

PRO TIP

When you have made your final repayment and all lodgements are up to date, a good tax agent will seek remission of penalties and interest applied.

CASE STUDY: Geoff and Lyn's nasty surprise

Prior to Next Level, Geoff and Lyn employed a bookkeeper. They didn't understand bookkeeping, and the lady they employed appeared to know her job.

After several years, she resigned suddenly. That's when Geoff and Lyn sought the advice of the team at Next Level. We performed an audit and clean-up of the file and found a nasty surprise. The bookkeeper hadn't processed quarterly superannuation payments for two and a half years!

That meant an $86,000 overdue superannuation obligation. When we showed Geoff and Lyn, they were distressed and angry, and reasonably so. They'd been paying the bookkeeper a solid wage to manage this issue.

The problem needed to be addressed. To correct this error, we prepared SG charge statements for each quarter the superannuation was missed. Fortunately, we managed to secure a twenty-four-month payment plan to settle the outstanding superannuation. The end bill was $99,500, up from the initial $86,000, because the tax office applied penalty interest and administrative charges on the overdue amount.

Geoff and Lyn, through sheer hard work, met the obligation. To commemorate the final payment, I splurged on a bottle of exquisite French champagne. Why? Because Geoff and Lyn had fixed their nasty surprise and had a $4,400 monthly boost to their cash flow because the debt was settled.

Now they could make money.

CASE STUDY: Mick's pleasant surprise

When Mick came to Next Level, he had overdue BAS payments totalling $102,000, with two more BAS returns outstanding. Once these two BAS returns were filed, the business's total debt was $160,000. This was not good news.

What did we do? The Next Level team organised a payment plan and incorporated the debt repayments into Mick's business cash-flow budget. This ensured that he was always able to make his payments on time.

Due to his exemplary compliance with the plan, once he had paid his debt, we were able to negotiate a remission of the interest that had been included in repayments. This meant that at the end of the payment plan, Mick's business received a refund of the $18,500 interest. A very pleasant surprise indeed!

Summary

In the Clean-up phase of the CLEAR method, you leave denial behind, face reality, become organised to tackle your problems head-on and look forward with confidence.

It's not easy, but it's worth it. I encourage anyone dealing with unresolved financial issues, and experiencing anxiety and stress due to unknown tax debt that questions the survival of their business, to take the necessary steps to resolve it and regain their freedom.

Always remember, sorting this out is achievable with the right plan in place. Get professional accounting advice as you'll never untangle the mess if you continue relying on the kind of DIY bookkeeping that led you there in the first place. A builder trying to fix their financial mess is like a skilled chef having a go at fixing a broken car engine on the way to work – both will get nowhere and end up with a headache.

Have confidence in your accountant. Be reassured that your tax bill won't mean jail if you communicate, lodge overdue returns and set up a plan to meet obligations. Once debt is cleared, you are on the way to a more predictable and profitable financial future.

You're now ready to move into the Learning phase of the CLEAR method where you begin a new journey. One that takes you to the enormous untapped potential lying doggo in your business.

SIX
Learn

With the Clean-up phase complete, you can be confident all tax obligations are met, your bookkeeping file is reconciled and debtor/creditor records are accurate. What lies ahead is the challenge to meet debt obligations while maintaining business profitability.

Where before bad choices created a mountain of accounting and finance problems, in the Learn phase, you'll see everything clear up. You start to grasp the overall picture of business finances, performance, position, margins and cash flow: the numbers.

When the right systems are in place, you can identify areas where money is being lost and rectify them. By strengthening financial controls to prevent profit leakage, you ensure your business perfor-mance improves. This will take six to twelve months, but towards the end, you will gain enjoyment and knowledge.

This chapter will show you how to accomplish this. At the end of this phase, you will be:

- **Bookkeeping savvy** – in the Learn phase, bookkeeping is done weekly by professionals

- **Owner wage driven** – focus is placed on paying the working owners a commercial wage, minimum $2,500 per week

- **Profit-driven** – your business will run at least at 10% operating profit

- **Targets aware** – you will know your business's daily targets to achieve 25/10

- **Tax planning efficient** – the business will, perhaps for the first time, be on top of future tax obligations

- **Monthly report savvy** – you'll have developed new skills in understanding your financial reports

Gain and pain in the Learn phase

Much in this phase hinges on the right accounting systems and processes as they allow you to track all business activity in real time. These systems provide you with the ability to observe the financial outcomes of decisions, which in turn builds confidence and accountability. This serves as a strong foundation for future development.

Understand, however, that during the initial stages of this phase, cash flow may be tight. This is because of the expenses involved in cleaning up your books and, in particular, resolving outstanding tax issues. Additionally, early on, your business finances may seem hazy, but with perseverance, you will overcome this and progress towards success.

CASE STUDY: Peter and Brett in the Learn phase

Diligent, hardworking and efficient in debt collection, Peter and Brett were successful in bringing cash money into their business but bills forever came out of left field. Every time they felt ready to expand and take on more staff, some other expense arose to take them back to square one. For them, with no grip on today, tomorrow looked difficult. Peter and Brett wondered what to do.

Over the dining table one night, Peter's wife wondered whether an accountant – a VCFO – with the right skills could help. That's when Peter gave Next Level a call.

At our first meeting, both men agreed there was big opportunity in the business. However, when we started talking about cash flow, Peter said, 'Our cash-flow problems keep hitting us like we're punching bags and with every blow, we feel more timid.'

Colourful language for sure, but I got the message. I heard them out and listed on a whiteboard the details of the Learn phase.

I asked, 'Are you in this phase?'

They nodded.

The financial visibility line

Financial visibility means to have clarity and understanding regarding a business's financial situation. It involves keeping accurate and up-to-date information on cash flow, revenue, expenses, profitability and performance. This information allows business owners to monitor their business's financial health, identify trends, track progress towards goals and make decisions that improve profit and performance.

The Learn phase of the CLEAR method is where this happens.

A whitepaper published by National Australia Bank found that 58% of small and medium-sized business owners recognised that sound financial management is a key success metric.[22] Guess what? To make that judgement, they all had to go through the Learn phase themselves.

Follow all steps of the Learn phase to cross over the financial visibility line – moving from murkiness to clear understanding of all your business numbers. Three to six months into the Learn phase, you will believe your business can achieve those financial goals you once dreamed about. You'll feel the clouds lift as you have financial visibility.

It's like gym training. You do push-ups, bench presses and squats. You stick to the regime and maybe approach exhaustion, only to find those first few weeks tend to mean much effort for little result. One day, though, the mirror reflects change – there's definition, muscle bulk. That's the visibility line. It's a huge moment.

With financial visibility, cash flow may remain tight, but it's being managed. Debts are not only met, but reduced, and you realise your business is bulking up.

Why? Because a sound financial plan is in place.

Six stages of the Learn phase

The six stages of the Learn phase are:

1. Outsource bookkeeping

2. Prepare a 25/10 budget

3. Work out the 25/10 breakeven point

4. Analyse monthly financial reports

5. Leverage your VCFO

6. Maximise tax savings

The Learn phase guarantees:

- You will know the required daily revenue and GPM to achieve your business's financial goals.

- You will have a clear understanding of your business's performance each month in relation to the financial goals you have set.

- You will know your expected tax liability in advance, the strategies you can implement and how much tax these strategies will save for your business.

Following all six stages of the Learn phase will reward you with these three guarantees, taking you from woebegone to winner. Let's see how Peter and Brett did it.

Stage One: Outsource bookkeeping

The Next Level team handled Peter and Brett's bookkeeping once a week using Xero software. This meant:

- **Timesaving.** Outsourcing allowed Peter and Brett to focus on core business activities. We recorded all transactions, performed account reconciliations and generated financial reports, which freed up their time to work on what they did well.

- **Accuracy**. We listed all transactions. This ensured that all financial information was entered, which eliminated risk of errors and discrepancies.

- **Cost-effectiveness**. Hiring a VCFO worked better than taking on a full-time employee. Peter and Brett saved on expenses such as employee benefits, training, computer and office space.

- **Peace of mind**. Knowing bookkeeping tasks were handled by a professional gave them mental space. Confident all records were organised, sequenced and regulation compliant, Peter and Brett could sleep well at night.

Stage Two: Prepare a 25/10 budget

After getting the bookkeeping right, we prepared a business budget to deliver two goals, which were – you guessed it – 25/10. Here's a copy of the annual budget report we prepared for them.

	INCOME	
	Sales	$3,960,000
(a)	**TOTAL INCOME:**	**$3,960,000**
	Less: COST OF GOODS SOLD	
	DIRECT COSTS	
	Materials & subs	$1,860,000

(b) TOTAL COST OF GOODS SOLD $1,860,000

(c) GROSS PROFIT [a - b] $2,100,000   (53%)

Less: Fixed Overheads

Advertising & Marketing	52,000
Computer Expenses	1,320
Consulting & Accounting	36,980
Electricity & Gas	4,580
Insurance	6,800
Merchant Fees	340
Motor Vehicle – Fuel	24,460
Motor Vehicle Leases	33,500
Motor Vehicle – Repairs & Maintenance	750
Postage	420
Printing & Stationery	3,380
Rent	68,429
Repairs & Maintenance	7,600
Staff Training & Development	3,500
Subscriptions & Memberships	23,450
Superannuation	113,011
Telephone	3,960
Tools Replacements	8,500
Uniforms	1,850
Wages – Trades	681,370
Wages – Admin	86,000

	Wages – Peter & Brett	260,000	
	Workcover	22,800	
(d)	**TOTAL FIXED COSTS**	**1,445,000**	
(e)	**OPERATING PROFIT** [c – d]	**655,000**	(16.5%)
	before interest, tax and depreciation		

> **PRO TIP**
>
> With your budget prepared like the one above, you will be able to calculate your breakeven point in under a minute.

Stage Three: Work out the 25/10 breakeven point

The breakeven point is the point at which total revenue equals total expenses. It represents the level of sales revenue and production capacity needed to cover costs and start making a profit. By understanding the breakeven point, owners can assess their business's performance, make informed decisions about pricing and production levels, and strive for profitability.

breakeven: total revenues = total expenses

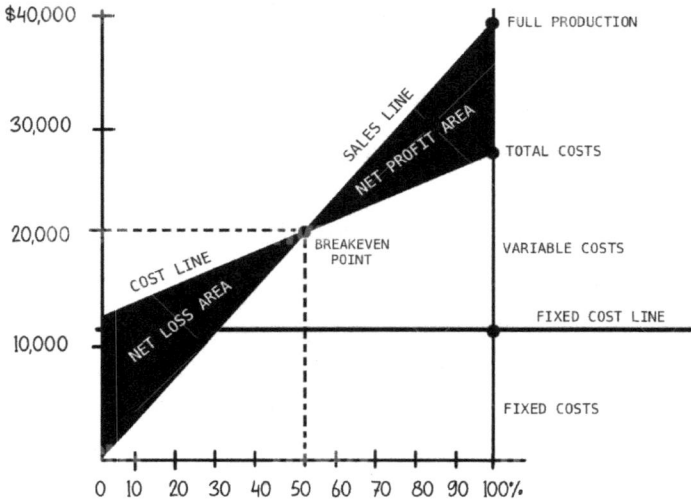

The diagram above shows breakeven point in the middle. This is where all costs have been covered, but no profit has been made. In this case, breakeven is achieved at $20,000 in sales revenues at a 50% production capacity. Any combination of sales/production below this level results in a loss, and any output higher means profit.

To calculate breakeven, use this formula:

breakeven = total fixed costs ÷ GPM

For example, Geoff and Lyn had monthly fixed operating expenses of $67,730 and a GPM of 86%. Their monthly breakeven point was $78,752:

breakeven: $67,730 ÷ 86% = $78,756

25/10 Breakeven point

The term '25/10 breakeven point' refers to the daily revenue required to achieve a 10% operating profit after paying the business owner(s) a wage of $2,500 per week.

CASE STUDY: Peter and Brett's 25/10 breakeven point

Like many people in business, Peter and Brett were not aware of how to calculate their 25/10 breakeven point. Once the team at Next Level showed them how to do this, it became a key pillar in shaping their business strategy.

Based on Peter and Brett's budget shown earlier in the chapter, to calculate the 25/10 breakeven point, they followed these steps:

(f) Annual breakeven = fixed costs **(d)** ÷ GPM **(c)**:

1,445,000 ÷ 53% = $2,726,415

(g) 10% of total income:

Total income **(a)** × 10% = $396,000

(h) 25/10 breakeven:

Annual breakeven **(f)** + 10% total income **(g)**
= $3,122,415

(i) Daily breakeven:

(h) ÷ 248 working days* in a year = $12,590

For Peter and Brett to reach 25/10, they needed $12,590 in daily revenue at a 53% GPM.

*How many working days in a year?

If the year is not a leap year, it has 365 days. Let's remove fifty-three Saturdays and fifty-two Sundays. It remains 260 days. Now remove twelve public holidays that do not fall on a weekend, and we obtain 248 working days in a year.

CASE STUDY: Learn phase guarantee #1 delivered for Peter and Brett

As revealed in Peter and Brett's annual budget, they were after 16.5% operating profit, not 10%, so the calculation became:

(a) Budgeted annual sales:

$$\$3,960,000 \div 248 \text{ working days} = \$15,968 \text{ per day}$$

Peter and Brett were now armed with crystal-clear financial visibility. They knew daily billable revenue of $15,968 was required at a 53% GPM to achieve budget. With this financial know-how, I am proud to say, they hit all targets by finding strategies to meet daily production revenue.

Peter said, 'We found our business mojo once we knew the numbers and the targets we had to reach.'

Stage Four: Analyse monthly financial reports

Driving from the Gold Coast to Sydney, you can follow road signs, but it's easier with GPS maps. Think of a budget as your financial GPS – it shows the easiest route.

The maps in this case are financial reports:

- Monthly profit-and-loss statements
- Monthly balance sheet
- Debtors summary sheets
- Creditors summary sheets

All reports are formatted to include the last three months and year-to-date totals.

When the Next Level team provided these reports to Peter and Brett, we emphasised the importance of them using a highlighter to mark any unclear items. Together, we went over the highlighted sections to explain the context behind the figures. While this process initially takes a bit of time, we have discovered through trial and error that it is the most effective method for business owners to learn to read and understand financial reports.

By developing knowledge about financial reports, you too can stay on track.

CASE STUDY: Learn phase guarantee #2 delivered for Peter and Brett – Clear understanding

Initially, Peter and Brett faced difficulties in comprehending the monthly financial reports, but by closely collaborating with their VCFO, they actively sought to improve their understanding. In our monthly meetings, they rapidly acquired the skills to comprehend and analyse the financial data, thanks to the valuable insights and expertise provided by the VCFO.

They made significant progress and gradually transitioned from being inexperienced to becoming proficient in interpreting the reports. This transformation was evident to both them and us.

How did we both recognise this change? Their questions evolved into confident statements of fact.

Stage Five: Leverage your VCFO

You now recognise that a solid knowledge of finances and accounting is essential to run a successful business, but if, like Peter and Brett, you're asking, 'How do I read financial reports? What's a balance sheet?' don't worry. You're not on your own. That's where VCFOs come in.

A VCFO is a high-performance accountant for owners that want their businesses to perform. They guide and fast-track your understanding of business finances to help you achieve your financial goals sooner.

Hiring a VCFO is an investment in both your business and yourself. Like a personal trainer at a gym, they know your goals, as well as the exercises and techniques to target areas that need developing, and they explain why you need to do a particular set of exercises.

The benefits of working out with the right VCFO are:

- **Financial analysis**. The VCFO checks the financial health and performance of the business, identifying areas for improvement and providing insights for informed decision making.

- **Budgeting and forecasting**. They develop recognised prediction methods to better plan and allocate resources.

- **Cash-flow management**. They monitor cash flow to ensure the business always has sufficient funds to meet obligations and seize opportunities.

- **Financial reporting**. They prepare and present accurate financial reports, including profit-and-loss statements, balance sheets and cash-flow statements.

- **Strategic financial planning**. They develop financial strategies. The VCFO's knowledge is your best asset when it comes to optimising profitability and managing financial risks.

- **Tax planning and compliance**. They investigate and recommend tax strategies to ensure compliance and maximise tax savings.

- **Business performance analysis**. They evaluate performance metrics against budget and recommend improvement strategies.

- **Cost control and expense management**. They identify cost-saving opportunities and implement cost control.

- **Scenario analysis**. They create models to simulate alternative financial scenarios and assess their potential impact.

- **Board and investor relations**. They provide reports for board meetings and investor communications.

If you want better or even exceptional financial results, a VCFO gives you the best shot.

Mastering the tax maze

A 2019 survey by the British Chamber of Commerce found 58% of small and medium business owners in the UK considered the tax system to be unfair.[23] This perception was even higher among microbusinesses, with 70% of owners believing the tax system to be unfair. Furthermore, 49% of small business owners said the government underestimated the time and money required to keep up with the complexities of an ever-changing tax system.

The message is clear – tax matters are best handled by a specialist who is licensed and stays updated with tax regulations. VCFOs fulfil this role as their business is to understand evolving tax legislation and its impact on clients. They navigate the laws and aim to maximise their benefits for you.

Stage Six: Tax planning maximises tax savings

Once you have a well-functioning system in place, which includes failsafe bookkeeping and a 25/10 budget, as well as a clear understanding of your 25/10 breakeven point, and hopefully, the assistance of a VCFO, you are now ready to maximise tax savings through tax planning.

Hats off to you, as you have already tackled the difficult part. The fruits of your effort will be seen in the form of significant tax savings generated through effective tax planning. These savings can amount to thousands or even tens of thousands of dollars each year.

What a VCFO examines:

- **Prepayments**. Some expenses can be made a year in advance to accelerate tax deductions. For example, Peter and Brett were able to prepay twelve months of rent in advance in June, thereby getting the tax deduction for the current year.

- **Materials**. A stocktake may be required to write off any obsolete or damaged stock.

- **Bad debts**. A review of the debtors' ledger to write off debts deemed uncollectible makes for a tax deduction.

- **Superannuation**. Often paying a little bit more into superannuation results in a personal income-tax deduction.

- **Business structure**. Family trusts are not always tax efficient. Small business restructure rollover provisions allow those eligible to restructure into a company, which means profit is taxed at a flat 25%.

CASE STUDY: Learn phase guarantee #3 delivered – a $22,000 tax holiday for Peter and Brett

After implementing the Learn phase stages, the team at Next Level put Peter and Brett in a position where there would be both a reduced tax burden and no more unexpected surprises. They now had clear financial visibility.

They worked hard to improve their business, and the results were good. So good, we anticipated an imminent tax problem. At our VCFO meetings, we designed a strategic tax plan to put them in front.

This is what we came up with. We recommended the business prepay $60,000 of office rent in June to cover the next financial year. This meant a $15,000 tax saving as a standalone strategy.

Next, we identified an opportunity for Brett to contribute $20,000 into his superannuation fund as a 'personal concessional contribution'. This made him eligible for a $20,000 deduction on his tax return, resulting in a saving of $7,800 in income tax. The combination of these strategies saved over $22,000 in tax.

Brett later said, 'Who knew taxes could be this much fun? It's a financial festival, where our VCFO makes sure we never pay a dollar more than necessary!'

The tickets to Bali were bought within a week.

Summary

At the beginning of this chapter, I mentioned that business owners often face cash-flow challenges when entering the Learn phase. However, once you navigate through this phase of the CLEAR method, the cash-flow difficulties will become a thing of the past. You will understand the importance of your numbers and margins by the preparation of your 25/10 budget and breakeven point and discover how effective planning can save you money. In other words, you've reclaimed control of your business numbers and finances.

By implementing the six stages of the Learn phase, you will gain clear financial visibility. This will enable you to make informed decisions and set yourself on the path to sustained growth and success.

Get ready for the next phase: Enjoy.

SEVEN
Enjoy

To enjoy means finding pleasure and satisfaction in the doing or experiencing of something. You will likely be at this point right now. You have cleaned up, organised your finances and learned.

This is the phase where you prosper and protect your business from financial uncertainty. It is consolidation, a chance to take satisfaction in how far you've come. You now have the time to enjoy business life.

Financially, your business is on a comfortable footing. Cash-flow cycles are regular and predictable. Gone is the monetary stress. Your expectations are being met and profits are matching or exceeding plan.

The initial pleasure in this phase comes from enormous relief. Once past that, you may push higher. If you are happy cruising, the Enjoy phase can endure, but should you have further ambition, my experience says you'll take a breather for six to eighteen months before pushing on.

Here are seven unmistakeable signs your business is in the Enjoy phase:

- **Cash flow**. The business can cover operating expenses, tax, superannuation and make profit.

- **Four bank accounts**. It uses specific purpose bank accounts.

- **Customer satisfaction**. It has a high level of customer satisfaction and loyalty.

- **Repeat business**. It has a strong base of repeat customers who continue to support its products or services.

- **Established reputation**. It has built a reputation for delivering high-quality products or services.

- **Strong customer retention**. It retains existing customers and has a low churn rate.

- **Continuous improvement**. It constantly seeks ways to improve.

Financial clouds clearing

In the Learn phase, you came up to speed with new processes, accounting and bookkeeping systems. Now you have emerged with a clarity of knowledge you likely once thought light years away. This alone does wonders for mental health and quality of life.

A 2020 report by McNair yellowSquares found Australian small business owners are more likely to be satisfied with life and feel happier if their businesses are established and stable.[24] Further, a 2022 report by the Australian social research consultancy firm Heartward Strategic found that there is a significant link between money and mental health.[25] People in difficulty are twice as likely to experience mental-health challenges than those who are not.

In the Enjoy phase, your business will be established and stable. At Next Level, we advise you now to set up a system of four bank accounts to ensure you always have the money to pay for everything. Cash is set aside into each account and always available when expenses are due.

Money with a purpose

Satisfaction and confidence come with the knowledge that your business operates on a workable plan. As the days, weeks and months tick by and your cash flow and profits improve, you may think about all the money in your bank account, but don't rush out and spend it. As money, like life itself, must have a purpose, here's what two leading financial sages have to say about it.

George S Clason, author of *The Richest Man in Babylon*,[26] which has been in print since 1926, advocates two key concepts:

1. Part of every dollar your business earns is yours to keep.

2. You should keep at least 10% of everything your business earns.

Mike Michalowicz offers a similar message in his book *Profit First*.[27] He says business owners should pay themselves first and let what remains dictate expense spending.

This way of thinking may be the opposite of your current operation. Previously, you likely paid all business expense commitments first, and then used that left over to pay yourself. Ring in the change; reverse that approach. Prioritise number one (yourself), then meet operating expenses.

How do you do this? To accomplish this, I recommend four bank accounts.

Four bank accounts

The four bank accounts your business needs are:

1. Income

2. Profit

3. Operating expenses

4. Tax

No trickery, just four bank accounts with four names. Each account is self-explanatory, although we will look at them in detail throughout this section.

The alternative – one big, jumbled account to handle income, profit, operating expenses and tax – is pointless. This is because it only provides a single bank balance for all purposes in your business. By separating funds into different accounts, you can monitor income, profit, operating expenses and tax obligations. It's specific accounts for specific purposes.

What could be simpler?

First step, direct 100% of your income into the income bank account. Then, twice a month, transfer a selected percentage to the other three accounts. As a guide, these percentages will be around:

- 10% to the profit bank account

- 60% to the operating expense bank account

- 30% to the tax bank account

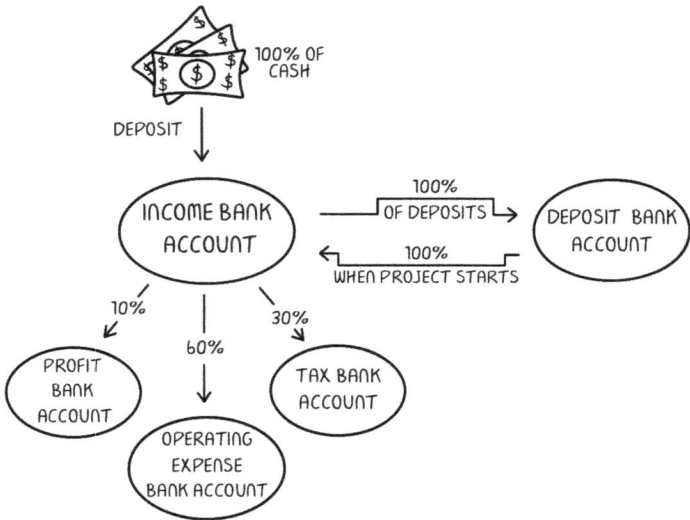

The 4/5 bank account system

Your percentages may differ. Ideally, you should determine your percentages with guidance from your VCFO and closely monitor workability over the next few months against your budget performance. Once operating, the bi-monthly transfers will give a good snapshot of your overall profit trend.

CASE STUDY: Geoff and Lyn's bank accounts

Geoff was working sixty hours a week and had one kidney operating at 40% capacity. He needed a transplant. A donor came up, but the operation would take him out of action for several months.

While he recovered, I worked with Lyn to set up four specific bank accounts. We made sure all customer payments were deposited into the bank account called 'Income'. Regularly, twice a month, first on the 15th and second on the last day of the month, 100% of that money was reallocated to the other three bank accounts – 10% to the profit account (remember: profit first), 60% to the operating expense account and 30% to the tax account.

To be clear, all money earned was set aside for profit, operating and tax – in that order.

It was the answer. The team at Next Level took it as a compliment when Lyn mentioned feeling 'right' about the set-up from the word go – female intuition is always on point.

Better news came – Geoff's transplant was successful, and over the next three months, we saw his strength improve out of sight. Upon his return to work, his energy levels were those of a man twenty years younger. Some said it was the operation, but I wondered whether it was due to him reading the profit figures!

The income bank account

The income account serves one and only one purpose: to collect 100% of payments from customers. It's a bucket; income goes into the account, and twice a month, it goes out to the three other accounts. This process is the foundation for the purposeful handling and provisioning of money. It is a precondition for success with the four bank account system.

The profit bank account

The first payment from the income account is to your profit account. If at first, you may be hesitant about 10%, start with 2%. Once you are satisfied about where this money is going, increase the allocation to 3%, then 4%. With each step, you are cultivating a profit-centric mindset, and one day, sooner than you may think, the allocation will be 10%. It's like a flower in full bloom!

This growth will serve as an enjoyable motivator for you to explore and implement strategies that further boost the percentage allocated to your profit account. Isn't that just what you have been waiting for?

The snowball effect

The snowball effect is a psychological term explaining how small actions cause bigger actions and often result in massive impact. Imagine a snowball rolling down Mount Everest. It gathers momentum, picks up more snow and, if conditions are right, becomes an avalanche. Your business has that possibility.

My goal is to get your 10% snowball rolling. However, at Next Level, we have often seen that there may be even greater potential. If achieving 15% or 20% allocation to the profit account becomes possible, seize all opportunities.

Profit account: Growth vs dividends

Business is like a tug of war. On one hand, the costs of growth, such as additional equipment, more staff, a bigger factory or a better website, pull you in one direction. On the other hand, external factors, such as the need to add another bedroom to your house, the expense of sending kids to a private school or the financial assistance required to support parents moving into retirement living, pull you in another.

What do you do? You find the line of best fit – you leave some money in the business and take some out.

Money taken for personal use is called a dividend, and best made according to a predetermined plan. This plan is referred to as a dividend policy, which we will discuss further in Chapter 9.

The operating expense bank account

This bank account pays for business expenses and some liabilities, including materials, subcontractors, wages, insurance, printing, stationery, telephone, rent, gas and electricity, repairs and maintenance, staff amenities and motor vehicle expenses. Lease and loan repayments are also made from this account.

Remember, one of your operating expenses is your own wage, and it's therefore drawn from the operating expense bank account. As discussed, you'll pay yourself $2,500 per week. The percentage that goes into this account is determined by your budget.

The operating expense account is like a rollercoaster ride. Every time you make a deposit (twice a month), it's all flush and abundant, but as expenses are paid and commitments are met, the account may hit a low point until the next round of money comes in. Don't worry, it's the nature of this account. Its balance goes up and down, and while it may feel strange at first, you'll get into the rhythm and embrace the ride.

The lower reaches of this up and down cycle serve as a hand brake on unnecessary discretionary spending. For example, a key supplier may say to you, 'Look, we've got a special on tiles this month. If you buy twenty boxes, we can give you a 5% discount.' Usually, you only buy a few boxes per month, but the discount sounds enticing. Should you take the deal?

The answer lies in the operating expense account. You'll know the allowance for monthly operating costs and the cash available and make your decision accordingly.

Most often, the decision is no. Sure, the discount may amount to $3,000, but if you don't need that specific colour tile, those twenty boxes could end up gathering dust in storage for years. Just like efficient money management, efficient stock throughput is equally important in building success.

With the checks and balances provided by the budget, you can see how actual operating expenses measure up. If there are major or consistent differences, you will need to re-evaluate the amounts allocated to better fit budget requirements.

When you have sufficient experience, the process will be quick and easy. In the end, it's all about making sure the account has the means to pay its dues and doesn't run out of money.

The tax bank account

I've suggested allocating 30% of the cash in the income bank account to the tax bank account. This will ensure that there is money readily available to meet tax obligations, whether they are monthly, quarterly or yearly.

The tax bank account pays for income tax, BAS and superannuation (see Chapter 5 for a detailed explanation of these terms in the 'Tax terms' section). The percentage allocated for this account will vary according to profitability and the tax bracket of your business structure.

The tax bank account permanently removes tax payment stress, but what if you find yourself in a pickle with it? Picture this: your business is booming, exceeding expectations and raking in profits. Amazing, right?

Here's the catch – if your tax percentage jumps to a higher bracket, you might be caught off guard. Suddenly, the money you saved for taxes falls short, leading to a frustrated juggling act with your finances. You don't want to fall into this trap!

Fear not, for your VCFO will spot the issue and help figure out the perfect percentage allocation. With their expertise, you'll never be caught short. I've seen it: a thriving business shows early signs of success in the financial year, which provides ample time to adjust the tax account and stay ahead of the game.

Sometimes things don't go as planned, and your business may underperform against budget, resulting in lower profitability. In such situations, you might find yourself allocating a higher percentage from your income bank account to your tax bank account than necessary. Decreasing the percent allocation to the tax bank account and increasing the percent allocation to the operating expenses account can free up some cash when times are temporarily tight. It can all be managed.

With the four bank account system implemented during the Enjoy phase, you will experience enjoyment from knowing your business finances are on track. As each month passes, your confidence in the method is likely to grow. The balances in each bank account are effectively supporting your business cash-flow needs.

Together, you and your VCFO will adjust the cash allocation needed to continue steering your business towards maximum performance. Isn't that the way to go?

CASE STUDY: Diagnosing Geoff and Lyn's financial problems

When Geoff and Lyn opened their four bank accounts, the Next Level team scheduled meetings to review their monthly and quarterly financial results against budget to keep everything on track. At a quarterly review meeting, telltale signs of financial headwinds emerged. Here's what we uncovered:

- The allocated percentage of money was not being transferred into the profit bank account, which called for a review. This analysis revealed that Geoff and Lyn purchased six months' worth of materials from a supplier, to take advantage of bulk order discounts. This depleted cash available to pay profit first.

- The operating expense account often ran out of money. We found they bought an $18,000 radio advertising campaign to increase sales – it didn't work and drained cash reserves.

- Their tax bank account was light on cash – this had been their best year ever with company profit soaring to $396,000. The percentage of income allocated to the tax account needed to be increased from 30% to 35%, thus rebalancing cash reserves to meet expected tax liabilities.

We pointed the issues out to Geoff and Lyn. Armed with this knowledge, they tackled the issues and successfully avoided long-term damage to their business.

The fifth bank account – deposits

I have deliberately separated this section on the deposits bank account from the four bank account system to avoid confusion.

A significant problem in the construction industry is the practice of taking a customer deposit and using it to fund the expenses of another project. This practice, commonly referred to as robbing Peter to pay Paul, can lead to a deteriorating financial situation, where your business relies on cash deposits from new projects to cover the expenses of current ones.

Setting up a deposit bank account keeps a clear line between separate jobs. When a customer's deposit is received, it must be transferred into the deposit bank account. Once that particular customer's project has started, the deposit is transferred back to the income bank account and allocated to the profit, operating expenses and tax bank accounts based on the specified percentages.

This process ensures transparency in cash flow and prevents the mixing of funds between projects. It not only helps accurately track customer deposits, but also provides a clear picture of the financial health of each project. Additionally, it simplifies management and ensures that expenses from one project are kept separate from deposits for another.

Summary

The Enjoy phase is where you experience real progress. Four bank accounts, plus one for deposits; more knowledge; better control and feedback on your finances – everything is on track.

I tell my clients to step back, celebrate, enjoy – they've earned it all. It is also about the time I suggest they ask themselves about the next phase.

So, dear Reader, do you have ambition?

EIGHT
Ambition

A t this point, everything is set for your business to skyrocket. During the Enjoy phase, you caught your breath while building up momentum, but now, you're ready.

Ambition is more than just a strong desire to achieve; it's a relentless pursuit fuelled by determination and hard work that propels you towards greater success and real accomplishment. In the Ambition phase, you turn aspiration into reality.

This chapter will reveal the secrets to boosting your business's net profit from 10% to 20% or even 25%. The duration of this phase depends on the magnitude of your goal. It could take anywhere from eighteen months to five years, or even longer if you're aiming for industry domination!

Here are six traits of a business in the Ambition phase:

1. **Visionary leadership**. The business has forward-thinking leaders who impart a clear vision for the company's future.

2. **Market opportunities**. Its leaders identify and capitalise on emerging market trends.

3. **Growth mindset**. Its leaders will take calculated risks in the pursuit of growth.

4. **Strong product or service differentiation**. The business offers unique products or services that set it apart from competitors.

5. **Talent acquisition and retention**. The business attracts and retains top talent.

6. **Strategic partnerships**. The business forms connections to access new markets, resources or expertise.

Seeds of ambition

In 2022, *Inside Small Business* reported 77% of Australians believe ambition is important, driving progress and innovative thinking.[28] This desire is set in the Enjoy phase. Ambition encourages creative thought on opportunity and advancement. It's about time to think, reflect and ask yourself:

- What's the vision for the business?

- Is there a big, maybe even audacious goal?

- Are there transformative decisions being avoided?

- Are there breakthroughs ahead?

- Are there inefficiencies within?

- Are there ways to gain more business?

- Is the business aligning with the future?

Remember, big thinking can set the vision you want to accomplish in the Ambition phase.

CASE STUDY: Geoff and Lyn's rework problem

In the Enjoy phase, Geoff and Lyn discovered a newfound sense of cash-flow confidence. Yet, although adequately allocating cash per the four bank account percentage plan, they had greater ambitions. They now wanted to push the profit percentage higher.

That's when the team at Next Level recommended a daily performance tracking chart. This measured, among other items, the amount of time spent reworking jobs. Rework refers to work needing to be redone because it was not done correctly the first time.

In Geoff and Lyn's case, the cost of rework was three times original cost. This was because the initial work was not up to scratch and required a second attempt. This brought about a third cost, which was lost income from the business being unable to take on a new job while the rework was done.

The chart below shows a black line for the hours of rework daily. At a glance, you can see the rework figure was too high.

Geoff said, 'Next Level uncovered one big area losing money – rework.' He and Lyn took the standout course of action – to get the job done right, first time. With that mindset, they set about improving their job systems. In a matter of a few weeks, the rework hours (black line) went down and the daily revenue (red line) went up.

Again, we go to Geoff. 'It was a complete business turnaround. Within the first three months, we made a full year's profit. The chart told us revenue was up, rework was down and production efficiency was on target. Hallelujah!'

Over a few months, Geoff and Lyn found a sense of financial swagger as the percentage allocated to the profit account increased to 18%. Over twelve months, the four bank account system was working a treat and their profit bank account had over $290,000 squirreled away. Plenty to both invest in growth and pay dividends.

'Time is money' is a saying that, for the first time, Geoff and Lyn understood.

You can acquire a version of the performance tracking chart by visiting my website and downloading it from https://nextlevelaccountants.com.au/resources/book-resources.

Time is money

The primary function of your business is to be profitable. Why provide services otherwise?

Time spent on work must be directly proportional to money made. Time is money, or more correctly, time equals money where trades and construction projects are concerned. If a project is delayed, it will lose money for these reasons:

- **Overheads** are fixed and remain constant regardless of changes in project schedules. Costs, such as rent, utilities and salaries, do not fluctuate. If a project experiences delay, it will result in additional overhead expenses. Because, for example, a week's delay is at least a week's more wages.

- **Cash flow**. Projects with a schedule overrun will, in effect, delay client payments, which in turn creates a shortfall in your business's cash flow. As a result, alternative means must be found to fund the cash-flow shortage.

- **Scheduling**. When projects' timelines are delayed, it keeps your trades, labour and equipment tied to the delayed project longer than anticipated. That is concerning when another project is due to start and resources are not available.

To avoid problems, your projects must run like a well-oiled machine. Be highly efficient and always on schedule.

Nailing down the numbers

Get your numbers wrong and life is a battle. Worse still, you could go broke. You must get your prices and methodology right.

It brings on the question: do you want to be busy in your business making money, or busy in your business not making money? Of the hundreds of businesses I've looked at, I can report that eight out of ten owners are busy not making money. If that's you, I say it's time to correct it. Let's nail the numbers.

I asked earlier, if you are a skilled tradesperson, how do you feel about financial management? Are you among the 81% of business owners who harbour doubt?[29]

It's not uncommon for entrepreneurs who doubt their abilities with financial management to be casual about reviewing and adjusting prices. Often, there is insufficient analysis, purpose or priority. If this sounds familiar, you must change it and propel yourself from the Ambition to the Reward phase.

The Next Level pricing methods have helped hundreds of clients improve profitability. These methods have been tested and consistently deliver results.

Project-management fee

According to SEEK, the average annual salary for construction project manager jobs in Australia ranges from \$130,000 to \$150,000.[30] Project management is an important responsibility in your business because it brings leadership and direction to projects. Without project management, your team is like a ship without a captain – moving, but without direction, control or purpose. Effective project management empowers team members to give their best work and deliver high levels of quality and service to customers. In many trades and construction businesses, the owner is the project manager.

The project manager:

- Oversees planning, scheduling and delivery
- Ensures project work is completed on time and within budget
- Organises logistics, delegates work and keeps track of job spending

This merits high pay, yet in my experience, many owners don't charge for their project-management time. Instead, they see it as an overhead, a cost of doing business. Not only is this mindset out of sync with best practice, but it is also guaranteed to cost you money.

Failing to charge a project-management fee is why many trades and construction businesses remain unprofitable or struggle to achieve desired operating profit. It is the largest contributing factor that hinders financial success, not charging your own time to a project.

As a VCFO, when I discover project-management time is not charged, or the time charged does not match actual project-management time, my immediate recommendation is to fix it, now. Set and charge the appropriate fee for project management. Align skills with profitability, not survival. If you have created significant value, the financial return must be commensurate.

Hourly charge-out rates

The second major area where I see trades and construction businesses fall short is in assessing the charge-out rate for their tradespeople. It is often underestimated.

When I press a business owner on how their charge-out rate works, the answer generally is, 'It's the industry standard.'

'But is it right?' I ask. Let's have a look.

Take a tradie working 38 hours a week. That's 38 hours × 52 weeks = 1,976 **(a)** hours in one year. If the tradie's wage rate is $36 per hour, that's $36 × 1,976 hours = $71,136 **(b)** per year.

Therefore, it costs $71,136 to pay the wage of one tradie for one year, right?

Wrong. Prices must factor in the extras required to keep them in the business. Here's a list of those costs.

Superannuation guarantee @ 11% of $71,136	$7,825
Payroll tax @2.4% > $650k	$1,707
Long-service leave @2.7%	$1,921
Work cover insurance @1.27%	$903
Training, trade school etc @1.5%	$1,607
Recruitment costs	$1,000
Other (eg phone, Christmas bonus etc)	$600
Total	**(c) $15,563**

It costs $15,563 on top of wages to keep that one employee in the business for one year. That is $71,136 + $15,563 = $86,699 **(d)**. This must be factored into the charge-out rate. Not only that, but annual leave, sick leave and public holidays must be factored in.

Annual leave (20 days × 7.6 hrs)	152
Public holidays (13 days × 7.6 hours)	98
Sick leave (10 days × 7.6 hours)	76
Total unproductive hours per year	**(e) 326**

While paid for 1,976 hours **(a)**, the tradie actually works 1,976 **(a)** – 326 **(e)** = 1,650 **(f)** hours. This means the true cost per hour of one tradie on full-time wages is $86,699 **(d)** ÷ 1,650 **(f)** hours = $52.50 per hour **(g)**.

To decide on actual job charge-out rate, we must factor in other expenses like factory rent, wages of administration staff, motor vehicle costs etc. Which, by the way, is not hard to do, as long as all costs have been recorded. Furthermore, we must factor in owner wages and company profit.

As a rough guide, the figure might be double the tradesperson's true cost per hour of $52.50 **(g)**: 2 × $52.50 = $105 per hour.

The point I make here is that every business owner needs to understand the true cost of labour in assessing business profitability. A VCFO can calculate hourly charge-out rates for any business and any trade, and apply that in calculating overall charge-out rates.

PRO TIP

Round the hourly charge-out rate up for simplicity. That means if you make the rate $105.15, round it up to $110 or more.

The charge-out rate that's right for your business

In the example above, the tradesperson's actual cost is $86,699, and your business generates twice their wages as revenue over the year, $173,398. When you have worked out the tradespeople's revenue contribution, you can use this number to calculate the labour revenue your business can make per year, based upon the number of tradespeople, their charge-out rates and the total this contributes to your business.

Each methodology has its merits. You may decide on a higher revenue contribution from tradespeople. For example, you may want to make $100,000 or $120,000 return on top of their wage.

Let's look at the numbers. Staff real wage cost is $86,699. Add $100,000 extra to the wage. This means the total return required per tradesperson is $186,699. Divide $186,699 by the 1,650 hours of actual work and the hourly charge-out rate is $113 per hour.

In the next scenario, staff real wage is $86,699. Add $120,000 extra to the wage. This means the total return required is $206,699. Divide $206,699 by the 1,650 hours of actual work and the hourly charge-out rate is $125 per hour.

Some trades and construction business owners purposely set rates. Unfortunately, many set their charge-out rates with a lick of the finger. They charge $80 because that's what their mate does in their business, and remember, the National Australia Bank survey says that is 81% of small and medium-sized enterprise owners.[31] Whew!

I argue every business owner must set rates based on actual figures. By doing this, you maintain necessary profitability.

In addition to understanding the true cost of labour, it is important for business owners to regularly review and adjust pricing strategy. Market conditions, competition and other factors can impact the product value for services. By regularly evaluating and adjusting your pricing, you ensure the appropriate fee is charged.

Be busy in your business, yes, but make money too.

Annual price reviews

Many things change over the course of a year. Labour shortages (or surpluses), tightening (or loosening) availability of materials, wage hikes, inflation – nothing stands still. Neither should your business.

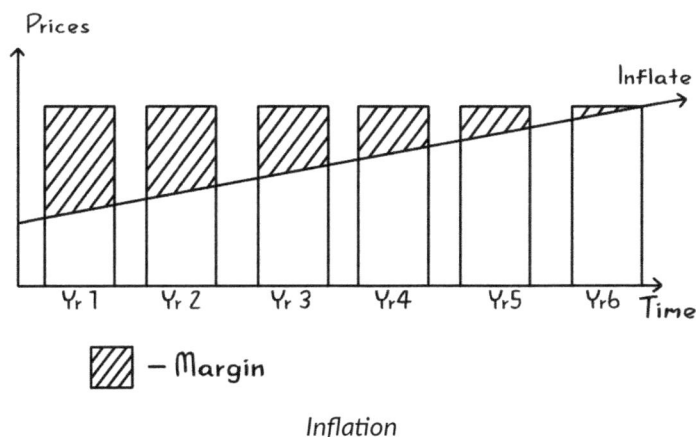

Inflation

To stay up with the pace of change, a business needs to undertake at least one review of the charge-out rates every year. If the environment is inflationary, necessitating price increases, there are various ways to go about it.

Let's look at an example. Business 1 and Business 2 are both in the same industry, their quality of work is equal, and their staff and abilities are on par. Business 1 conducts annual price reviews by analysing the numbers and understanding the market, and decides to increase prices by 5% each year, starting from 2019. Here are its charge-out rates:

Year	Charge-out rate
2019	$100
2020	$105

Year	Charge-out rate
2021	$110
2022	$115
2023	$121
2024	$127

Business 2, on the other hand, does not conduct annual price reviews, and for five years has kept the charge-out rates fixed. In 2024, upon realising income isn't stretching as far as it used to, the owner jacks up the price 5% per year covering the last five years, 25% in total.

Its figures read:

Year	Charge-out rate
2019	$100
2020	$100
2021	$100
2022	$100
2023	$100
2024	$125

See the difference? In 2024, Business 2 had to get a $25 per hour increase past existing customers. A big ask.

Wait, there's more! From 2020 to 2023, as costs soared, profits took a nosedive. Sure, Business 2 could have attracted more customers with its lower charge-out rate compared to Business 1, but it would come at the expense of increased pressure and demands on its staff. To add insult to injury, in 2024, its charge-out rate was still lower than its rival's. Not exactly a recipe for success, is it?

What does this tell you about the need for annual price reviews? If your business ambitions are high, then annual price reviews are the method to get you there.

I might add, many of my clients review prices every three to six months, and most volume builders review every month. If the sharpest in the business do it, how about you?

PRO TIP

Do your annual price reviews when setting the budget for the year ahead.

Classifying clients to increase prices

If a business hasn't increased its prices for five years (and there are a lot of them out there), this can be seen as a tough challenge. Although it's easy with new customers, the difficulty lies in increasing prices with your existing clients.

When it comes to repricing for your existing clients, classifying them into A, B and C works well:

- **A clients** – highly profitable

- **B clients** – seem profitable, but you're not too sure

- **C clients** – unprofitable; you are not making any money from them

Start with the C clients, the ones you're losing money on. If you let them go, you're creating more space for A clients.

To increase your prices by 25%, come clean with clients and say, 'Look, it's our fault. We haven't been charging you the correct rate for the last five years, and unfortunately, we can't continue on that basis. We need to increase our prices by X.' Blame me if you must!

Increasing prices can seem daunting. I know, I've been there. When I did this in my business, I found that nineteen customers out of every twenty accepted the new prices. I even had a couple of clients say, 'I'm surprised you didn't do this sooner', and you know what? The one in twenty I lost, great. In most cases, they were a pain.

Your business should not carry losers.

Next, start chipping away at the B clients. From experience, I can say you will get a fair hearing and the majority will accept your proposals. By this time, you will have the skills and confidence to handle what can be a hard phone call to make.

Now you will be ready for your A clients. Guess what? A clients usually have long-time experience at making similar calls themselves, so they are the most understanding.

PRO TIP

Never communicate price increases by sending an email. Doing so risks the relationship. It must be done in person or over the phone. Have a methodical approach. Start with the worst C client and work up from there to your best A client.

The contract variation money hole

Project variations can be a headache for trades and construction businesses. Contract variations create an administrative burden, increase project risks and affect profitability. To make a profit, you must implement a standardised procedure. Otherwise, variations will become a sinkhole, swallowing your profits.

These are the ways to lose through contract variations:

- **Productivity reduction** – labour, equipment and materials are diverted from the original contract work, resulting in lost time and money.

- **Increased estimating time** – this is not an overhead. Someone (perhaps you) must prepare the variation. Not charging for the time cost results in lower profit margin.

- **Incidental haste** – as variations delay production output, time pressures often mean the overall schedule is not thought through. Mistakes can be monumental.

How do you get it right?

The variation administration charge (VAC)

1. Add a VAC to all agreements

The VAC offsets the time needed to adjust production schedules and properly evaluate the scope of work and materials necessary to complete the variation. The amount you charge depends on the size of your business, your industry and the average length of your trades projects.

Charge a flat fee for a VAC, perhaps $350, and work up or down from there. This charge is fair for the disruption caused and discourages unnecessary or frivolous requests.

It is important for the charge to be documented in contractual agreements. Highlight it to your clients. Tell them, 'I'd like to draw your attention to the VAC', and explain what it is and why it is charged. By so doing, you ensure transparency and maintain a professional customer relationship. That way, misunderstandings are avoided.

By all means, discuss variations like window sizes during a project, but if clients request a formal quote for new materials, labour and scheduling, a VAC must be administered.

2. Have a standard price formula

A standard price formula for all variations ensures a level playing field. This allows your team (not just you) to handle variations following a simple profitable process.

The formula could look like this:

- $350 VAC

- 100% markup on materials

- 20%, 30% or 50% labour markup

- 50% markup on project-management time

The reason for the time markup is the tricky nature of contract variations. Any change causes delay and often leads to issues you may not have thought likely. As a professional, you must allow for that possibility.

Follow these two steps and turn variations from profit traps to profit generators.

Scenario planning

To ensure long-term profitability, your business needs to be agile and able to adapt to changing market conditions. Running scenario planning sessions helps a business maintain the agility to act and adapt.

In the ceaseless march of time, opportunities spring forth and setbacks occur. Such is the rhythm of life. Being prepared in the present moment is the strongest weapon if you wish to seize control of the unfolding events and turn the tide in your favour.

A 2022–23 survey by CPA Australia found that 47.6% of businesses had grown in the previous twelve months and that 55.1% expected to do so in the following twelve months.[32] Whether you are part of the 55.1% or not doesn't matter. I say to 100% of you, 'Give yourself the challenge!' Be ready to take opportunity when it arrives. Plan for alternative scenarios.

How do you do that?

Scenario planning workshop

A good starting point is a half-day workshop. You supply the tea and cucumber sandwiches while the VCFO brings the whiteboard.

Watch and encourage everyone to throw in ideas on:

- Anticipated changes, particularly in relation to revenue, GPMs, overheads and employment
- Winning future work and how it might relate to capacity and cash flow
- Possible financial obstacles
- Developing a strategy

Bold? Yes. Imaginative? Yes.

If you're serious about hitting the next level, it's the way to go.

Keep in mind, nothing is pie in the sky, nothing is sacred. No matter how unorthodox or wild the suggestion, it's open for discussion. You never know what it may flush out.

One crazy idea

Someone once told me a story about overground electric power lines going down under heavy snow in the USA. At a scenario planning discussion,

> somebody suggested renting a helicopter to
> fly directly over the power lines to blow off the
> accumulated snow using the downdraught of the
> helicopter blades.[33]
>
> This crazy idea worked!

The point to this anecdote is to evaluate all scenarios, no matter how preposterous, for they may flush out the very idea needed to propel your business forward. Let the discussion unfold. Select the most promising ones.

Your VCFO can create a three-year budget in accordance with targets defined from your scenario planning workshop. Once your detailed financial forecasts are constructed, you will have valuable insights into potential risks and opportunities and be better equipped to make informed strategic decisions.

CASE STUDY: Geoff and Lyn's two-part business

There were two parts to Geoff and Lyn's business – factory manufacturing and offsite work. After becoming more familiar with finances, they asked the team at Next Level to look at their operations.

The figures indicated the factory was efficient, profitable and, according to Geoff, smooth to run.

'But offsite?' I asked.

'Yes, it makes money,' answered Geoff, 'but it sure takes up a lot of our time.'

We examined the books and ran the numbers. It did, indeed, make money, but when brought back to return per hour worked, the offsite came up short.

When we asked Geoff and Lyn about that, it seemed offsite not only took up much of their time, but due to labour shortages, the scheduling was constantly thrown out of order. The 'issue', they confessed, 'Occupies our thoughts 24/7. The stress is colossal.'

'So,' we asked, 'if you expanded the factory, would you get more work?'

Without thinking about it, they both said yes.

That's the way they went. The offsite was sold and factory capacity increased. Now not only are their returns greater than ever, but they get to spend weekends on their boat, listening to the footy while fishing.

Summary

You're through the Ambition phase. It's happening. You're on top of your operations, prices and finances. You're experiencing the pleasure of knowing your business is running right. It's a mighty achievement.

Sure, there are challenges ahead, but your business is now set up in the way to handle them best. While business is not easy, it's probably fair to say it's less difficult and you have become a master, not a servant, of your time.

Congratulations – you have made it!

What now?

NINE

Reward

The Reward phase is where the investment of time, money and effort in your business peaks. It is the fruition of your belief and achievement, and rightful compensation for your hard work over the journey.

Having made a good living, you may one day consider the many options about how to go forward. You might choose between continuing to work hands-on in the business every day or winding down. You might wish to bring your children or trusted colleagues on board. On the other hand, you may seek new opportunities, greater work-life balance or to retire.

Above all, Reward is the phase that causes my clients to reflect.

Here are seven traits of a business in the Reward phase:

- Profitability is consistent
- Cash flow is healthy
- The owner's involvement is 100% by choice
- Customer base is loyal and growing
- Brand recognition is strong
- Market leadership is respected
- Employee morale is high

If your business ticks these boxes, I will wager you are successful.

Business structure

Success necessitates a structure that aligns fluent levels of involvement, control and responsibility. If your business becomes so large that management, office and site staff don't have enough time to keep up, you need to consider corporatisation. That means establishing a CEO and board.

Let's define these two terms:

- **The CEO** runs day-to-day management and is responsible for profits or losses. Subordinate managers and staff answer to the CEO; hiring and firing are at their discretion. Other terms often used to describe the same role are executive director, managing director, general manager and chief supervisor.

- **Boards** oversee the management and direction of a company. They make strategic decisions, set policies and ensure the company meets objectives. They have a fiduciary duty to protect the interests of the company's shareholders.

Who is higher, the CEO or the board? While the board appoints, evaluates and can fire the CEO, the latter holds the higher position.

Overall, it's the responsibility of both the CEO and the board to move the business forward. In deciding whether to go the CEO/board route, the business owner needs to know whether they still want to lead and in what capacity – as CEO, board member or perhaps chairperson of the board. It's a big decision, but big rewards await.

Here's an illustration of what such a CEO/board organisation would look like:

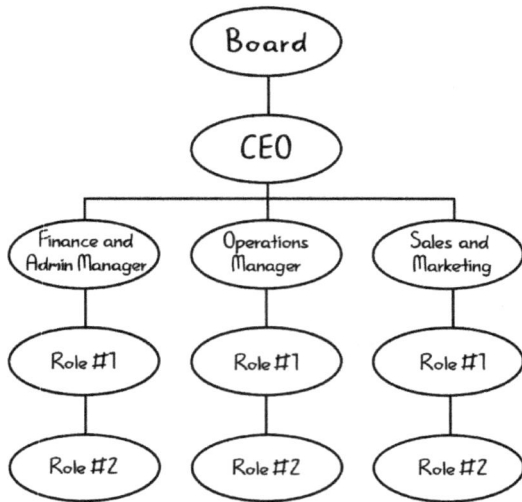

The CEO/board structure

CASE STUDY: Geoff and Lyn's Reward phase

When Geoff and Lyn made the Reward phase, their business was exceeding budget. Target achievement was on course, so they found the time at last to enjoy life outside the business.

Geoff dived headfirst into his passions, amassing a collection of six classic cars, a fishing boat and two jet skis. With his car club, he could drive to Lake Eildon or Corio Bay and fish in the serene waters. Lyn was able to spend more time supporting her daughter with a disability and was able to help her son set up his first home.

How did this happen? Geoff and Lyn appointed a CEO to manage the business. While they still visit for a few hours most days, they aren't actively managing the business themselves.

It makes sense. After twenty years of hard yakka – dealing with customers, staff and the tax office, while coping with existing deadlines and trying to gain new work – it was time to ease back. They did exactly that.

In fact, Geoff once said to me, 'My mobile isn't always on these days, so if you want to get in touch, try a letter marked "Post restante, Corio Bay".'

I haven't written that letter yet, but one day I might.

Dividend policies

A dividend policy outlines how the company will distribute its dividends to shareholders (business owners). It will provide details on:

- How often dividends are paid out: monthly, quarterly or annually

- How the amount of the dividend will be calculated

A dividend policy can provide several benefits for a private business. Firstly, it allows the business to distribute profits to owners/shareholders, providing them with a return on their investment. Secondly, it contributes to the overall financial management of the business. By setting a clear dividend policy, the company can allocate funds appropriately, balancing the distribution of profits with the need for reinvestment and business growth.

Additionally, a dividend policy improves owner satisfaction. Regular dividend payments help maintain a positive relationship between owners and managers. In summary, a well-designed dividend policy provides financial benefits, attracts investors, enhances owner / manager relationships and supports effective financial management for a private business.

There are two dividend policies to consider: residual and stable.

Residual dividend policy

The residual dividend policy distributes a fixed percentage (such as 30%) of the balance in the company's profit bank account during each dividend. This ensures that while the dividend amount will fluctuate based on profitability, a significant percentage of the cash profit always remains in the business's profit bank account.

Let's look at an example where there is a 30% residual dividend policy, paid out quarterly.

Bank balance 1 July		*$150,000*
July profit	$45,000	
August profit	$47,500	
September profit	$45,000	
Total Quarter 1 profit	$137,500	+$150,000
	Total	**$287,500**
Less 30% dividend payment (ie 30% of 287,500)	Dividend #1	−$86,250
Bank balance 1 October		$201,250
October profit	$50,000	
November profit	$60,000	
December profit	$62,000	
Total Quarter 2 profit	$172,000	+$201,250
	Total	**$373,250**
Less 30% dividend payment (ie 30% of 373,250)	Dividend #2	−$111,975
Bank balance 1 January		$261,275
January profit	$25,000	
February profit	$30,000	
March profit	$55,000	
Total Quarter 3 profit	$110,000	+$261,275
	Total	**$371,275**
Less 30% dividend payment (ie 30% of 371,250)	Dividend #3	−$111,380

Bank balance 1 April		$259,895
April profit	$42,000	
May profit	$47,500	
June profit	$50,000	
Total Quarter 4 profit	$139,500	+$259,895
	Total	**$399,395**
Less 30% dividend payment (ie 30% of 399,395)	Dividend #4	−$119,820
Bank balance 30 June		*$279,575*

This example shows that over a period of twelve months, the business:

- Paid four quarterly cash dividends to owners, totalling $429,425

- Grew the cash balance in the profit bank account from $150,000 to $279,575, an 86% increase

Could this policy run your profit account down to empty? No. As long as the business generates profit and has the right systems in place, it is as failsafe as possible.

Stable dividend policy

This policy provides owners / shareholders with a fixed dividend, regardless of company performance. The goal in this case is to align dividend policy with long-term growth, and as such, the set payment amount is adjusted according to the owner's assessment of future performance.

Profitability is identical to the previous example, but the stable dividend policy is to pay a fixed $120,000 to shareholders every quarter. With a stable dividend payout of $120,000, the profit bank account now might read:

Bank balance 30 June		$150,000
July profit	$45,000	
August profit	$47,500	
September profit	$45,000	
Total Quarter 1 profit	$137,500	+$150,000
	Total	**$287,500**
Less dividend payment		−$120,000
Bank balance 30 September		$167,500
October profit	$50,000	
November profit	$60,000	
December profit	$62,000	
Total Quarter 2 profit	$172,000	+$167,500
	Total	**$339,500**
Less dividend payment		−$120,000
Bank balance 31 December		$219,500
January profit	$25,000	
February profit	$30,000	
March profit	$55,000	

Total Quarter 3 profit	$110,000	+$219,500
	Total	$329,500
Less dividend payment		−$120,000
Bank balance 31 March		$209,500
April profit	$42,000	
May profit	$47,500	
June profit	$50,000	
Total Quarter 4 profit	$139,500	+$209,500
	Total	**$349,000**
Less dividend payment		−$120,000
Bank balance 30 June		*$229,000*

This example shows that over a period of twelve months, the business:

- Paid four quarterly cash dividends to owners, totalling $480,000

- Grew the cash balance in the profit bank account from $150,000 to $229,000, a 53% increase.

Could this policy run your profit account down to empty? Potentially, yes, but this dividend payment's settings of policy would be determined by management's assessment of future performance. It is unlikely the profit bank account would be reduced to nil.

In the Reward phase, you may consider giving back to the community that has provided your business with opportunities. You make a living by what you receive, but a life by what you give. If one day you decide your business is about more than just the bottom line, a community giving policy is something you may look at.

How much is your business worth?

Another key element of the Reward phase is to calculate how much your business is worth. This is done by combining goodwill value with business assets. Together, these represent a fair reflection of the work you've put in and profits generated. The value of the business becomes a big consideration when you as the owner wish to move on.

Goodwill

Goodwill refers to intangible assets such as brand recognition, customer loyalty and the reputation that the company has. It accumulates over the course of a business's journey and has significant impact on the business's value.

For example, take a local hamburger shop. It's fair to say this will have low goodwill compared to a

McDonald's store due to the latter's extensive branding, reputation and consistent product.

Let's look at a business that has generated consistent profits over twenty years, with the last five years bringing in steady annual profit of $250,000. Standard practice is to calculate goodwill by multiplying one year's profit by two (known as the earnings multiple). In this case, $250,000 multiplied by two equals a goodwill value of $500,000. If stock, plant and equipment are valued at $1,500,000, the total value of the business is $500,000 (goodwill) plus $1,500,000 (assets), which equals $2million.

The actual process is more complex than the simple example above, as the earnings multiple varies due to many factors. It reflects the level of risk associated with goodwill. A lower perceived risk results in a higher earnings multiple, while higher perceived risk leads to a lower earnings multiple. A helpful way to understand this is that buyers will pay a higher earnings multiple (ie a higher price) for a business with lower risk. With your input, a good VCFO will work out the value of your business.

Below are earnings multiple statistics derived from *The Messy Marketplace*,[34] written by Brent Beshore, founder and CEO of a private equity firm. These statistics tend to be a reasonable guide.

Business earnings before interest, taxes, depreciation and amortisation	Earnings multiple
$500k–$1m	× 2.5–3.5
$1m–$2m	× 2.75–3.75
$2m–$5m	× 3.5–6.5
$5m–$10m	× 4.5–8

CASE STUDY: The value of Mick's business

Mick wanted out.

To calculate his operating profit, the team at Next Level took net profit and removed interest, depreciation and amortisation, which are non-operating expenses. Without sounding over complicated, we made a 'weighted' profit calculation based on the previous three years. This weighted calculation settled on 60% of last year's profit (see below), plus 25% from two years ago and 15% from three years ago, combining these to achieve the weighted earnings. The profit calculated was $650,000.

We then applied an earnings multiple to the weighted profit ($650,000) to estimate goodwill. In this case, considering the size of and risk associated with Mick's business, an earnings multiple factor of three was reasonable.

Mick's goodwill calculation was $650,000 × 3 = $1,950,000.

Finally, we added fair market value for stock, plant and equipment, furniture and fittings and other assets.

Here's the calculation:

Year	Earnings	Weighting	$s
2023	705,000	× 60%	423,000
2022	625,000	× 25%	156,250
2021	471,000	× 15%	70,650
Total weighted earnings			$650,000 *(rounded up)*
Times earnings multiplier			× 3
Goodwill value			$1,950,000
Add assets			
Add stock			$20,000
Add plant, equipment, furniture etc			$215,000
Total value of Mick's business			**$2,185,000**

Planning for exit

Exiting a business is usually done in one of three ways:

1. Outright sale on the open market

2. Succession passing on to a trusted colleague

3. Legacy: placing in the hands of your children

At Next Level, we have found that the choice is not a hard one to make as the owner has enough life experience to come to the most suitable decision. For the record, though, it is never too early to plan your exit.

PRO TIP

Consult with your trusted VCFO advisor to grasp the key considerations such as business valuation, buy-in options, tax implications and legal obligations.

Business owners choose to exit their business for a variety of reasons, some of which are:

- Retirement
- Health issues
- Exhaustion
- New business opportunities
- Lifestyle changes
- Receiving an attractive offer

How to make the exit

- **Management buyout**. Key staff members may wish to purchase the business outright or take a controlling interest. This may be done by means of vendor finance. That is, the seller lends or part lends the money to the buyer for the purchase. To accept this, the exiting vendor must believe in the buyer's ability to pay off the loan, but if they have worked alongside the buyer for many years, they'll have an idea of the risk level.

- **Carve-out**. If your company operates from several branches or properties, there are at least two businesses: the property and the operational business. Each one has a different rate of return, along with varying risks and value. An owner might choose to sell each part separately. For instance, they might sell the operating company while retaining the property to earn rental income.

- **Staged exit**. This strategy involves selling a minority interest, also known as a partial sale. It is effective when both the seller and buyer aim for a gradual transfer of ownership.

- **Private equity**. There might be people interested in purchasing your business, particularly during consolidation or rationalisation in your industry. Private equity can facilitate a management buyout, provide a phased exit for you as business owner or propose an outright purchase.

- **Initial public offering** for share market listing. In this case, your business needs to be large and have a suitable governance structure with a compelling growth story. For only that will attract institutional investors to underwrite the listing and keep the share price solid. While this can be expensive, it provides liquidity to owners. The business is listed on the stock exchange, so the owner can sell their shares at the click of a button.

- **Wind up**. If a buyer cannot be found, the final option is to wind up the business by selling its assets. This is generally undesirable as any potential goodwill value is forfeited, but in the face of an immediate crisis, you may need to consider this option. While not the best solution, it's always available.

This emphasises the importance of planning for your business exit. Unforeseen circumstances such as a health crisis can occur, and without preparations, you will be caught off guard.

PRO TIP

Milk the metaphorical cow. To maximise business value prior to sale, have a strategy to boost profits, reduce costs, improve efficiency and enhance goodwill. For example, profit boosting could mean exploring productivity improvement measures or expanding the customer base. Cost reduction involves renegotiating supplier contracts or reducing overheads. Efficiency improvement includes streamlining processes and/or implementing new technologies. Enhancing goodwill refers to strengthening the brand or improving customer relationships. All these can significantly increase business value, but require careful planning and execution. Work with your VCFO to milk your cow.

CASE STUDY: Succession – Mick's staged exit

Mick's construction business entered the Reward phase. Jane, a friend and employee of fifteen years, was appointed CEO to oversee business operations. Mick, as owner, continued to serve as a chairman/ board member.

While still providing guidance, Mick now had the freedom to enjoy his life outside of the business. In fact, he went on a family tour of Europe, during which he took the opportunity and recognised the need for self-reflection. Upon his return, Mick approached Next Level to create a business succession plan.

From the options we presented, he chose a staged exit plan over three years, which involved gradually selling 50% of his business equity to Jane. In other words, over three years, 50% of the value of his business – $1,092,500 (see previous case study) – would be paid to Mick in cash. As each part payment was received, an equivalent percentage of equity was transferred to Jane. Upon completion of the sale, Mick's remaining equity entitled him to an estimated $210,000 a year in dividends. Moreover, his fee as chairman drew $80,000 a year.

Mick had a great outcome, receiving $1,092,500 in cash as a return on his share capital. Additionally, he retained a non-executive role and received an annual $290,000 in income through director fees and dividends.

Similarly, Jane had a great outcome as she bought into a business she loved, trusted and felt secure in. This business will serve as a financial plan for her family's future.

Legacy: Selling to the kids

This is in a category of its own because when making the transfer to their children, owners don't always apply the necessary due diligence. The first order of the day is to get a business valuation, then set up a sale arrangement that has legal weight.

Why? Sometimes, kids expect parents will hand the business over for free, and that, inevitably, leads to all sorts of issues – particularly if the business goes backwards. Yes, gifting is an option, but an exiting owner needs to be in an otherwise secure financial position.

Capital gains tax (CGT) concessions

Small business CGT concessions can cut or even eliminate capital gain from your business's goodwill value. However, certain eligibility requirements must be met to seize these powerful tax breaks, and some concessions may be accessible only if the sale corresponds with your retirement.

The four concessions are:

1. Fifteen-year exemption

2. Small business 50% reduction

3. Small business rollover

4. Small business retirement exemption

Be aware that utilising concessions could eliminate or significantly reduce CGT. Engaging your VCFO to strategically prepare your tax plan before a sale leaves no stone unturned in the search for tax savings.

CASE STUDY: Geoff and Lyn's CGT concessions

Geoff and Lyn decided to sell their business. Starting from scratch, they had painstakingly built a solid powder-coating business over twenty years and had amassed goodwill to the value of $1.75 million. The looming question was how much CGT would have to be paid?

After analysing eligibility factors, the Next Level team determined Geoff and Lyn could claim a small business CGT concession, the most significant one being the fifteen-year exemption. This concession meant no CGT to pay. In celebration of their triumph, Geoff and Lyn embarked on a luxurious round-the-world cruise, a well-deserved reward for their dedication and effort over many years.

What a fitting finale. Congratulations, Geoff and Lyn!

Summary

This Reward phase completes the CLEAR process: you will have financial security and can make the call on your level of business involvement. From low or no engagement to high-level structures involving boards and CEOs, it's up to you.

Either way, I hope you enjoy the ride!

Conclusion

D rawing from over two decades of experience as a trusted advisor and accountant, interacting with almost 1,000 business owners, I've seen first-hand how 80% grapple with financial difficulties. It's with these insights that I penned this book, aiming to provide a transformative guide to assist business owners in overcoming financial adversity, reclaiming control and charting their path to financial success using the CLEAR method.

The most rewarding aspect of my role as a VCFO is working with tradespeople who have self-awareness, understand their knowledge gaps and are not afraid to acknowledge what they don't know. This helps position them to invest in the resources required to improve profitability settings. They all comprehend that any business cannot grow without profit. With the right advice, profit becomes the guiding light in all their decisions and strategies.

Business owners in trades and construction frequently make five major financial mistakes. The first is doing their own bookkeeping. While it may seem cost-effective, it invariably leads to errors and inefficiencies that cost more in the long run.

The second mistake is not having a financial budget. Without a clear budget, businesses often overspend in some areas and underspend in others.

The third mistake is not knowing the breakeven point. This is where income equals cost. Understanding this means better-informed decisions on pricing and sales strategies.

The fourth is not reviewing financial reports regularly. Reports provide insights into financial health; neglecting them leads to missed opportunities or unseen risks.

The final mistake is not planning for taxes. This usually results in a business paying more tax than necessary.

It is important to know all these mistakes are avoidable. Overlook one and you may well struggle and find the stress and anxiety overwhelming, because suddenly the realisation that the business is larger than you hits home.

Business is an ecosystem that many people depend on. Foremost is your family. Financial returns not matching efforts will cause tension at home. Then come staff, subcontractors, customers and the community who also rely on you. Plain sailing is only possible with profit. Losses let everyone down. By setting and meeting a firm financial goal (think 25/10), you navigate the way through all pressure.

A primary obstacle many owners have in reaching financial goals lies in self-perception. They often see themselves as tradespeople first, owners second. Switch the perspective: be an owner/tradesperson. It gives you a better handle on staff, equipment and processes, for only then can exceptional service and craftsmanship be the calling card that customers remember.

A big – maybe the biggest – obstacle remover is the budget. It's an indispensable tool, and like any tool, its results are dependent on the expertise of the user. High-quality output goes hand-in-hand with high expertise. In a sense, your VCFO is also a tool. By arming you with in-depth analysis of your financial history, business overheads and margins, your VCFO provides powerful insights and can make strategic recommendations to further raise your level of expertise as business owner.

The team at Next Level has empowered hundreds of trade and construction business owners to triumph over struggle and reach a level of success where they can proudly say, 'I've made it.' All we did was engage our CLEAR method.

Let me touch on the points again:

- Clean-up

- Learn

- Enjoy

- Ambition

- Reward

Clean-up means tidying up tax compliance, streamlining bookkeeping and establishing payment plans for outstanding tax debt.

Learn illuminates the path to implement proper accounting and finance systems in your business. The goal? Unparalleled financial visibility. You'll have daily revenue targets, breakeven points, tax liabilities insights and monthly performance updates vs financial goals. This isn't just achieving financial visibility – it's stepping over the threshold into a world of financial clarity and control.

Enjoy is the moment that the fog of confusion and stress dissipates, replaced by financial clarity. Debts to the tax office vanish. Cash flow is not just flowing but pouring. It's the time to rejoice in the results of all your hard work and create a financial fortress heading into the future.

Ambition means if the sky is the limit, you might want to position yourself in the stratosphere. If you aim high up, an exciting ride lies ahead.

The **Reward** phase is about introspection – looking within. The business is mature, so your involvement becomes optional. Will you appoint a full-time CEO, retire or part retire, sell or pass the business on? The choice is yours. No matter: the successful owner is able to make that choice from a position of financial strength. With the CLEAR method and the help of a trusted VCFO, that's you.

To those who have honoured me by reading this book, I hope it has reset your thoughts. Running a business is no easy task. In the end, if you're going to be in business, you must always stay ready to take opportunity as it presents.

Good luck!

Notes

1   Australian Small Business and Family Enterprise
    Ombudsman, *Small Business Matters* (ASBFEO,
    2023), www.asbfeo.gov.au/sites/default/
    files/2023-06/Small%20Business%20Matters_
    June%202023.pdf, accessed 15 February 2024
2   National Australia Bank, *Moments that Matter:
    Understanding Australian small to medium
    businesses* (NAB, 2017), https://business.
    nab.com.au/wp-content/uploads/2017/06/
    J002580_MTM-Whitepaper-IPSOS-v4_C1.pdf,
    accessed 15 February 2024
3   G Gilfillan, 'Definitions and data sources for
    small business in Australia: A quick guide'

(Parliament of Australia, 2015), www.aph.gov.au/
about_parliament/parliamentary_departments/
parliamentary_library/pubs/rp/rp1516/quick_
guides/data, accessed 20 January 2024

4   G Stephensen, 'Measuring business success'
(Lloyds Corporate Brokers, no date), www.
lloydsbrokers.com.au/Measuring-Business-
Success.htm, accessed 11 January 2024

5   C Waters, 'Half of small business owners earn
less than minimum wage, report finds', *Sydney
Morning Herald* (30 August 2019), www.smh.
com.au/business/small-business/half-of-small-
business-owners-earn-less-than-minimum-
wage-report-finds-20190830-p52me8.html,
accessed 11 January 2024

6   M Guta, '82% of business failure is due to poor
cash management (INFOGRAPHIC)', *Small
Business Trends* (2019), www.smallbiztrends.com,
accessed 11 January 2024

7   S Roddy, 'How small businesses budget'
(Clutch, 2021), https://clutch.co/accounting/
resources/why-small-businesses-need-budgets,
accessed 11 January 2024

8   J Petty, 'Start me up!' (University of Technology
Sydney, 2005), www.uts.edu.au/sites/default/
files/Start_me_up.pdf, accessed 11 January
2024

9   K Airs, '"How much extra did you pay the
government last year?" MP's post praising
Kerry Packer's famous speech about the
"stupidity of legislation, politics and tax"

divides the internet', *Daily Mail* (2021), www.
dailymail.co.uk/news/article-9949339/MP-
celebrating-Kerry-Packer-speech-stupidity-
legislation-politics-tax-divides-internet.html,
accessed 22 January 2024

10 S Musgrave, 'How do Australian small
businesses measure success? Not the way you
might think' (Nine Advisory, 8 November 2018),
www.nineadvisory.com/blog/small-business-
success, accessed January 2024

11 A Turner-Cohen, '"Wiped out": "Horrific"
claim about Australia's collapsing construction
industry', *News.com* (19 June 2022), www.news.
com.au/finance/business/manufacturing/
wiped-out-horrific-claim-about-australias-
collapsing-construction-industry/news-story/
b808556d41ab8cb29fd9a3f90b2e3c07, accessed 11
January 2024

12 S Danckert, S Johanson, N Sambul, J Gordon
and R Clun, 'Building industry "on the brink"
after two groups collapse in 24 hours', *Sydney
Morning Herald* (31 March 2023), www.smh.
com.au/business/companies/home-builder-
porter-davis-in-trouble-20230330-p5cwvn.html,
accessed 11 January 2024

13 ME Gerber, *The E-Myth Revisited: Why most
small businesses don't work and what to do about it*
(Harper Business, 2001)

14 J Ashford, *Selling to Serve: Sell your accounting
and bookkeeping services with unshakeable confidence
for more than you thought possible* (independently
published, 2021)

15 E Connolly, D Norman, T West, *Small Business: An economic overview* (Reserve Bank of Australia, 2012), www.rba.gov.au/publications/workshops/other/small-bus-fin-roundtable-2012/pdf/01-overview.pdf, accessed 20 January 2024

16 P Campbell, 'Hidden costs of discounting' (ProfitWell, 2018), www.mahavajiralongkorn.com/discounting-benchmarks-for-subscription-companies.html?wtime=92s, accessed 11 January 2024

17 Heartward Strategic, *Money and Mental Health: Social research report* (Heartward Strategic, 2022), www.beyondblue.org.au/docs/default-source/about-beyond-blue/20061-money-and-mental-health-research-final-report-220804.pdf?sfvrsn=fd5d30e5_2, accessed 11 January 2024

18 McNair yellowSquares, *Small Business and Mental Health: Supporting Small Business when they are Facing Challenges: Report prepared for Department of Industry, Science, Energy and Resources* (McNair yellowSquares, 2020), www.industry.gov.au/sites/default/files/2021-01/small-business-owners-and-mental-health-report.pdf, accessed 11 January 2024

19 S Roddy, 'How small businesses budget' (Clutch, 12 May 2021), https://clutch.co/accounting/resources/why-small-businesses-need-budgets, accessed 11 January 2024

20 G Stephensen, 'Measuring business success' (Lloyds Corporate Brokers, no date),

www.lloydsbrokers.com.au/Measuring-Business-Success.htm, accessed 11 January 2024

21 Score Association, 'New infographic: The burden of small business accounting, taxes and payroll', *PR Newswire* (28 January 2015), www.prnewswire.com/news-releases/new-infographic-the-burden-of-small-business-accounting-taxes-and-payroll-300026479.html, accessed 11 January 2024

22 S Musgrave, 'How do Australian small businesses measure success? Not the way you might think' (Nine Advisory, 8 November 2018), www.nineadvisory.com/blog/small-business-success, accessed January 2024

23 Bedfordshire Chamber of Commerce, 'BCC: Small businesses don't see UK tax system as level playing field' (23 April 2019), www.chamber-business.com/blog/bcc-small-businesses-dont-see-uk-tax-system-as-level-playing-field, accessed 11 January 2024

24 McNair yellowSquares, *Small Business and Mental Health: Supporting Small Business when they are Facing Challenges: Report prepared for Department of Industry, Science, Energy and Resources* (McNair yellowSquares, 2020), www.industry.gov.au/sites/default/files/2021-01/small-business-owners-and-mental-health-report.pdf, accessed 11 January 2024

25 Heartward Strategic, *Money and Mental Health: Social research report* (Heartward Strategic,

2022), www.beyondblue.org.au/docs/default-source/about-beyond-blue/20061-money-and-mental-health-research-final-report-220804.pdf?sfvrsn=fd5d30e5_2, accessed 11 January 2024

26 GS Clason, *The Richest Man in Babylon: The success secrets of the ancients* (Originally published 1926; revised edition Signet, 2004)

27 M Michalowicz, *Profit First: Transform your business from a cash-eating monster to a money-making machine* (Portfolio, 2017)

28 B Smart, 'Why small businesses need to protect big ambition', *Inside Small Business* (29 March 2022), https://insidesmallbusiness.com.au/management/growth/why-small-businesses-need-to-protect-big-ambition, accessed 11 January 2024

29 G Stephensen, 'Measuring business success' (Lloyds Corporate Brokers, no date), www.lloydsbrokers.com.au/Measuring-Business-Success.htm, accessed 11 January 2024

30 What can I earn as a project manager? (SEEK, January 2024), www.seek.com.au/career-advice/role/project-manager/salary, accessed 19 January 2024

31 S Musgrave, 'How do Australian small businesses measure success? Not the way you might expect' (Nine Advisory, 8 November 2018), www.nineadvisory.com/blog/small-business-success, accessed January 2024

32 CPA Australia, *Asia-Pacific Small Business Survey 2022–2023* (CPA Australia, 2023),

www.cpaaustralia.com.au/-/media/project/
cpa/corporate/documents/tools-and-
resources/business-management/small-
business-survey/australia-market-summary-
2022-23.pdf, accessed 11 January 2024

33 N Culver, 'Power companies bring in
helicopters to clear snow', *The Spokesman-Review*
(26 December 2015), www.spokesman.com/
stories/2015/dec/26/power-companies-bring-
in-helicopters-to-clear-snow, accessed 4 March
2024

34 B Beshore, *The Messy Marketplace: Selling your
business in a world of imperfect buyers* (Boring
Books, 2018)

Further Reading

Ashford, J, *Selling to Serve: Sell your accounting and bookkeeping services with unshakeable confidence for more than you thought possible* (Independently published, 2021)

Clason, GS, *The Richest Man in Babylon: The success secrets of the ancients* (Originally published 1926; revised edition Signet, 2004)

Holdaway, B, *Make Money Simple Again: Financial peace in less than 10 minutes a month* (Major Street, 2022)

Michalowicz, M, *Profit First: Transform your business from a cash-eating monster to a money-making machine* (Portfolio, 2017)

Priestley, D, *24 Assets: Create a digital, scalable, valuable and fun business that will thrive in a fast changing world* (Rethink, 2017)

Tracy, B, *No Excuses!: The power of self-discipline* (Vanguard, 2011)

Wickman, G, *Traction: Get a grip on your business* (BenBella Books, 2012)

Ziglar, Z, *Born to Win: Find your success code* (Prabhat Prakashan, 2021)

Acknowledgements

I would like to thank the amazing group of humans that have selflessly supported me with their wisdom, insights and guidance. I owe you all a huge debt of gratitude for everything you have done over the years.

In writing the book, I would especially like to acknowledge Simon Russell (aka Big Si) for being beta reader, coach and mentor. Your attention to detail and suggestions were a large part in the book's overall success.

To my parents, Peter and Pauline Russell: growing up with you and watching you sparked my passion for small business. Peter, thank you for mentoring me as an accountant in my formative years and teaching me what you know.

I'd like to thank James Ashford for his advice on how to get the book started and Peter Georgiadis for being a mentor and beta reader.

The Author

Lyndon Russell leads Next Level Accountants. He resides in Torquay on the Surf Coast, Victoria, with his wife Carolyn and their two teenage children. His passion for entrepreneurship and business was sparked by observing his parents successfully run multiple businesses in hospitality, retail and professional services.

Lyndon is an author, qualified accountant and tax agent. He is a Fellow of the Institute of Public Accountants, a Fellow Financial Accountant, holds a master's degree in professional accounting and has been a trusted advisor to businesses for over twenty years.

The CLEAR system, created by Lyndon, is a proprietary five-step method that has helped hundreds of businesses to quickly gain financial visibility and achieve prosperity, enjoyment and reward. This ensures that business owners are paid their worth, and their businesses generate a net profit exceeding 10%.

Lyndon works alongside an amazing team of people at Next Level, pursuing the company mission to become a leading provider of accounting services to trades and construction businesses in Australia. In his spare time, he indulges his passion in the surf lifesaving community and sits as director and treasurer of the Jan Juc Surf Life Saving Club.

You can get in touch with Lyndon and find out more about his work here:

- ⊕ www.nextlevelaccountants.com.au
- 🔗 www.linkedin.com/in/lyndon-russell-8881369
- 🅕 www.facebook.com/nextlevelaccountantsAUS

www.ingramcontent.com/pod-product-compliance
Lightning Source LLC
Chambersburg PA
CBHW071545200326

41519CB00021BB/6617